中国互联网发展报告
2017

中国网络空间研究院　编著

電子工業出版社
Publishing House of Electronics Industry
北京•BEIJING

内 容 简 介

本书是世界互联网大会蓝皮书的重要组成部分，全面反映了中国互联网发展成就、发展现状和发展趋势，系统地总结了中国互联网发展的主要经验，深度分析了中国在网络基础设施、信息技术、数字经济、电子政务、网络安全、网络空间国际治理等方面的战略规划、政策举措、发展成效、发展水平和未来趋势；开创性地构建了以六大核心要素为基础的中国互联网发展指标体系，对全国各省域互联网发展情况进行综合评价和排名，对准确地把握我国互联网发展水平具有重要意义。

本书汇聚了国内外互联网领域最新研究成果，使用最新案例和权威数据，内容丰富、重点突出，对政府管理部门、互联网企业、科研机构、高校等互联网领域从业人员全面了解和掌握世界互联网发展情况具有重要参考价值。

图书在版编目（CIP）数据

中国互联网发展报告. 2017 / 中国网络空间研究院编著. —北京：电子工业出版社，2018.3
ISBN 978-7-121-33357-6

Ⅰ. ①中…　Ⅱ. ①中…　Ⅲ. ①互联网络－研究报告－中国－2017　Ⅳ. ①TP393.4

中国版本图书馆 CIP 数据核字（2017）第 321889 号

策划编辑：郭穗娟　齐　岳
责任编辑：郭穗娟
印　　刷：北京画中画印刷有限公司
装　　订：北京画中画印刷有限公司
出版发行：电子工业出版社
　　　　　北京市海淀区万寿路 173 信箱　邮编　100036
开　　本：720×1000　1/16　印张：15.75　字数：250 千字
版　　次：2018 年 3 月第 1 版
印　　次：2018 年 3 月第 1 次印刷
定　　价：180.00 元

凡所购买电子工业出版社图书有缺损问题，请向购买书店调换。若书店售缺，请与本社发行部联系，联系及邮购电话：（010）88254888，88258888。

质量投诉请发邮件至 zlts@phei.com.cn，盗版侵权举报请发邮件至 dbqq@phei.com.cn。

本书咨询联系方式：（010）88254502，guosj@phei.com.cn。

前　言

当前，以信息技术为代表的新一轮科技和产业革命正在兴起，为经济社会发展注入了强劲动力。中国互联网把握历史机遇，应势而动，顺势而为，进入发展快车道，经过23年波澜壮阔的发展，已成为举世瞩目的网络大国，探索出了一条具有中国特色的互联网发展治理之路，为世界互联网发展做出了中国贡献，创造了中国经验。

在这一大背景下，我们启动了《中国互联网发展报告2017》（以下简称《报告》）的编撰工作，力求详尽深入地对中国互联网发展进行研究和分析，总结历史、剖析现状、引领未来，使报告成为了解中国互联网历史、读懂中国互联网现状、把握中国互联网趋势的最好读本。为此，《报告》做了三方面的努力。

1. 始终把习近平总书记网络强国战略思想作为报告编写的理论基础和根本遵循

党的十八大以来，习近平总书记高瞻远瞩、把握大势，紧密结合我国互联网发展治理实践，提出一系列新理念新思想新战略，系统阐述事关网络安全和信息化发展的重大理论和实践问题，形成内涵丰富、博大精深、科学系统的网络强国战略思想，成为习近平新时代中国特色社会主义思想的重要组成部分。特别是党的十九大提出要建设网络强国、数字中国、智慧社会，推动互联网、大数据、人工智能和实体经济深度融合，发展数字经济、共享经济，培育新增长点、形成新动能，为新时代网信事业发展指明了前进方向、提供了根本遵循。我们将总书记网络强国战略思想作为报告的灵魂和主线贯穿通篇，深入宣传阐释总书记治网理念、思想和主张，以帮助广大读者全面领会、准确把握总书记网络强国战略思想的丰富内容和精神实质。

2. 始终把中国互联网发展的生动实践作为报告编写的研究基础和现实依据

在习近平新时代中国特色社会主义思想特别是网络强国战略思想指引下，中国互联网发展治理取得重大成就，网络空间日渐清朗，国家网络安全屏障进一步巩固，信息化驱动引领经济社会发展作用凸显，网络空间国际话语权和影响力明显提升，人民群众在共享互联网发展成果上有了更多获得感。这些生动实践为报告编写工作提供了广阔的空间和丰富的资源。《报告》在重点阐述 2017 年我国互联网发展现状，突出新实践、新成就的同时，对中国接入互联网 23 年来特别是党的十八大以来我国互联网发展情况进行了全面回顾，是对互联网治理中国经验的系统展现。

3. 始终把全面、准确、客观作为报告编写的学术目标和工作准则

在报告编写过程中，我们始终坚持突出国际性、权威性、精准性，强化思想性、理论性、概括性，体现出中国学界对互联网发展态势的一种认识和思考。我们坚持内容的全面性，对我国网络基础设施、信息技术、数字经济、电子政务、网络安全、网络空间国际治理等方面进行全领域、全景式研究；坚持数据的准确性，综合权威性、时效性等因素，择优采集官方、行业、研究机构三大类数据来源；坚持评价的客观性，在借鉴国内外成熟指标体系的基础上，设立并发布中国互联网发展指数指标体系，从六大维度对全国 31 个省（自治区、直辖市）的网络安全和信息化工作进行综合评估，以全面准确反映全国及各地互联网发展水平。

我们期待，《报告》的编写和出版将为中国互联网发展提供新动力，为世界了解中国互联网提供新窗口，为世界推动互联网发展提供新选择。这也是《报告》编撰工作的初衷和使命！

中国网络空间研究院
2017 年 12 月

目　录

总 论

纵观世界文明史，人类先后经历了农业革命、工业革命、信息革命。每一次产业技术革命，都给人类生产生活带来巨大而深刻的影响。当今世界，社会信息化和经济全球化相互促进，以互联网为代表的信息技术日新月异，引领了社会生产新变革，创造了人类生活新空间，拓展了国家治理新领域，极大提高了人类认识世界、改造世界的能力。中国正处在信息化快速发展的历史进程之中。中国党和政府高度重视互联网发展，自 23 年前接入国际互联网以来，按照积极利用、科学发展、依法管理、确保安全的方针，加强信息基础设施建设，发展数字经济，推进信息惠民。

党的十八大以来，以习近平同志为核心的党中央举旗定向、运筹帷幄，迎难而上、开拓进取，革故鼎新、励精图治，统筹国内国际两个大局，统筹推进“五位一体”总体布局，协调推进“四个全面”战略布局，坚定不移贯彻落实新发展理念，坚定不移推进改革攻坚，坚定不移推进党风廉政建设和反腐败斗争，解决了许多长期想解决而没有解决的难题，办成了许多过去想办而没有办成的大事，推动党和国家事业发生历史性变革，确立了习近平新时代中国特色社会主义思想的指导地位。党中央高度重视网络安全和信息化工作，把完善互联网管理领导体制作为十八届三中全会确立的 60 项改革任务之一，成立中央网络安全和信息化领导小组，统筹协调政治、经济、文化、社会、军事等各领域的网络安全和信息化重大问题。习近平总书记高瞻远瞩、把握大势，紧密结合我国互联网发展治理实践，提出一系列新理念新思想新战略，系统阐述事关网络安全和信息化发展的重大理论和实践问题，形成内涵丰富、博大精深、科学系统的网络强国战略思想。党的十九大提出新“两步走”战略，开启了中国特色社会主义新时代，特别是对推动互联网、大数据、人工智能、共享经济发展，建设网络强国、数字中国、智慧社会等作出重大战略部署，为中国互联网发展提供了根本遵循与行动指南。

一、中国互联网发展成就

在习近平新时代中国特色社会主义思想特别是网络强国战略思想指引下，中国互联网发展治理的顶层设计和总体架构基本确立，网上正能量更强劲、主旋律更高昂，网络空间日渐清朗，国家网络安全屏障进一步巩固，信息化驱动引领经济社会发展作用凸显，网络空间国际话语权和影响力明显提升，人民群众在共享互联网发展成果上有了更多获得感，网信事业发展取得重大成就。

1. 网络基础设施成为推动经济社会发展的新型公用基础设施

习近平总书记指出，我们要加强信息基础设施建设，强化信息资源深度整合，打通经济社会发展的信息“大动脉”。中国把握全球互联网发展趋势，深入实施“宽带中国”战略，开展网络“提速降费”行动，持续推进网络基础设施建设和演进升级，宽带接入能力、网络服务质量和应用水平显著提升，宽带网络规模和建设水平国际领先。截至 2017 年 6 月，中国骨干网带宽已超过 100Tbps,国际互联网出入口带宽达到 7974.779Gbps，互联网宽带接入端口数量达到 7.39 亿个，固定宽带用户数超过 3.22 亿户，移动通信用户总规模超过 13.6 亿户，其中 4G 用户达 8.9 亿户[1]。农村信息基础设施实现跨越式发展，电信普遍服务能力大幅提升，实现了 100%的乡镇和 93.5%的行政村通宽带。应用基础设施快速发展，内容分发网络（CDN）节点数超过 2 400 个，峰值宽带储备超过 10TB。互联网关键资源拥有量大幅增长，IPv4 地址数量 3.38 亿个，IPv6 地址数量 21 283 块/32 地址，均居世界第二位。中国(.cn)域名注册数达到 2 061 万个[2]，成为世界第一大国家顶级域名。

2. 网络信息技术取得重大突破

习近平总书记指出，要加快信息领域核心技术自主创新，着眼抢占信息技术发展制高点，与发达国家差距不断缩小，个别领域实现全球领先。通过实施创新驱动发展战略，推进一批关系国家全局和长远的重大科技项目，集中力量开展互联网关键共性技术、前沿引领技术研究，在高性能计算、量子通信、人工智能、云计算、大数据、核心芯片、操作系统、5G 等前沿技术方面取得重大突破。截至 2016 年年底，中国计算机、通信和其他电子设备制造业有效发明专利数 227 365 个，2016 年专利申请数 118 725 个，近五年中四度位居全球国际专利申请量第一位。中国首颗量子科学实验卫星“墨子”号成功上天。全球首条量子保密通信网络“京沪干线”全线贯通，覆盖距离超过 2 000 km。中国自行研制的全球卫星导航系统北斗卫星导航系统实现规模化应用，首次发射两颗北斗三号卫星，进入全球组网新时代。积极参与国际通信技术标准化制定，主导完成了多项 IETF 标准，在 IPv4/IPv6 互通、网络安全、路由协议等领域实现了重点突破，对下一代互联网核心技术的影响力逐步增强。

---

1 数据来源：工业和信息化部，《2017 年 1—6 月份通信业经济运行情况》；CNNIC，第 40 次《中国互联网络发展状况统计报告》。

2 数据来源：CNNIC，第 40 次《中国互联网络发展状况统计报告》。

3．以数字经济为代表的新经济蓬勃发展

习近平总书记指出，网信事业代表着新的生产力、新的发展方向，应该也能够在践行新发展理念上先行一步。中国充分发挥信息化对经济社会发展的驱动引领作用，大力发展数字经济，在电子商务、信息服务、网络终端设备、互联网金融、共享经济及“互联网+”融合等领域取得突出成效。2016 年，中国数字经济规模总量达到 22.58 万亿元人民币，跃居全球第二，占 GDP 比重达 30.3%[3]，成为经济增长的新引擎和新亮点。电子商务作为当前中国数字经济最活跃的领域，2016 年网络购物用户数达 4.67 亿人，电子商务交易额达 26.1 万亿元[4]，网络零售交易额达 5.2 万亿元，居全球首位。共享经济呈井喷式发展态势，2016 年市场交易额达 3.45 万亿元[5]。互联网金融快速发展，2016 年电子支付业务量达到 1 395.6 亿笔，业务金额达 2494.5 万亿元[6]。中国坚持鼓励支持和规范发展并行，政策引导和依法管理并举，为网信企业发展营造良好市场环境，阿里巴巴、腾讯、百度、华为等一批网信企业国际影响力日益提升，大量具有创新活力的互联网企业纷纷涌现。网信企业对接资本市场发展成效显著，截至 2017 年 10 月底，网信领域上市公司数量达 730 家，总市值达到 17.5 万亿元人民币，在全球互联网企业市值前 20 强中，中国企业占据 8 席。

4．电子政务发展创新社会治理

习近平总书记强调，要加快推进电子政务，鼓励各级政府部门打破信息

---

3 数据来源：中国信息通信研究院，《中国数字经济发展白皮书（2017 年）》。

4 数据来源：商务部，《中国电子商务报告 2016》。

5 数据来源：国家信息中心分享经济研究中心、中国互联网协会分享经济工作委员会，《中国分享经济发展报告（2017）》。

6 数据来源：人民银行《支付体系运行总体情况》。

壁垒、提升服务效率，让百姓少跑腿、信息多跑路，解决“办事难、办事慢、办事繁”的问题。中国将“互联网+政务服务”列入《“十三五”国家信息化规划》的 12 项优先行动，相继出台了《促进电子政务协调发展的指导意见》《关于加快推进“互联网+政务服务”工作的指导意见》等政策。各地按照中央统一要求部署，持续深化电子政务应用，建立政务网络互联、信息共享、业务互通机制，推动公共服务一体化在线服务平台建设，提供跨部门、跨领域的全流程、“一站式”在线服务。截至 2016 年年底，全国 31 个省（自治区、直辖市）及新疆生产建设兵团中，27 个地区已经建立省级网上政务服务平台[7]。政务新媒体快速发展，截至 2017 年 6 月，经过新浪平台认证的政务微博账号达到 17 万个，政务类微信公众账号达 51.4 万个，成为推进信息公开、加强政民互动、畅通民主监督的新渠道，有力保障了公民知情权、参与权、表达权和监督权。

5. 网络安全保障能力不断提升

习近平总书记指出，没有网络安全就没有国家安全，没有信息化就没有现代化。安全是发展的前提，发展是安全的保障。中国制定出台《网络安全法》《国家网络空间安全战略》等一系列法律法规和战略规划，建立健全网络安全责任制、网络安全审查机制、网络安全监测预警响应机制，修订《国家网络安全事件应急预案》，提升网络安全态势感知、事件分析、追踪溯源以及遭受攻击后的快速恢复能力。强化关键信息基础设施安全保护，首次在金融、能源、电力、通信、交通等重要行业领域开展全国范围的关键信息基础设施网络安全检查工作，以查促改、以查促建。加快网络安全标准化建设，截至 2016 年年底，全国信标委发布信息安全国家标准数量达 195 个[8]。网络

7 数据来源：《电子政务蓝皮书：中国电子政务发展报告（2017）》，社会科学文学出版社。

8 数据来源：赛迪智库。

安全产业规模迅速扩大，网络安全企业超过 1 000 家，2017 年网络安全产业规模有望超过 1355 亿元[9]。建立网络安全一级学科，公布 7 个一流网络安全学院建设示范项目，21 所高校成立网络空间安全学院。坚持网络安全为人民、网络安全靠人民，强化网络安全宣传教育，连续举办 4 届国家网络安全宣传周活动，覆盖人数超过 20 亿人次，全社会网络安全防护意识和技能明显提升。

6．网络空间日渐清朗

习近平总书记指出，网络空间是亿万民众共同的精神家园，网络空间天朗气清、生态良好，符合人民利益。网络内容建设与管理始终坚持正能量是总要求、管得住是硬道理，加强网上正面宣传，创新传播手段，把握网上舆论引导的时、度、效，提高新闻舆论的传播力、引导力、影响力、公信力，推出了一大批符合网络传播特点、群众喜闻乐见的现象级新闻报道和主题宣传产品，有力提升了正面宣传报道的到达率、阅读率、点赞率。大力培育和践行社会主义核心价值观,实施中华优秀文化网上传播工程，培育积极健康向上的网络文化，提升网民网络素养，用社会主义核心价值观凝聚人心、滋养社会。综合运用法律、行政、技术等手段，全面加强网络生态治理，完善网信执法体制机制，推动依法管网、依法办网、依法上网，强化主管部门属地管理责任、网站主体责任，提高工作科学化、规范化水平。充分发挥社会组织、企业、专家学者等各方面作用，建立多主体协同参与的网络生态治理模式。完善违法违规信息和网站快速联动处置机制，围绕网民反映强烈的问题，完善举报工作机制，2016 年，受理举报近 4 000 万件次，有害信息有效处置率达 92%以上。深入开展“清朗”“护苗”等 20 余个专项行动，有效处置各类违法有害信息，为人民群众营造天朗气清的网络空间。

---

9 数据来源：中国电子信息产业发展研究院，《2016—2017 年中国网络安全发展蓝皮书》。

7. 网络空间国际话语权和影响力不断增强

习近平总书记在多个重要国际场合阐述国际治网理念，特别是提出推进全球互联网治理体系变革“四项原则”和构建网络空间命运共同体“五点主张”，为推进全球互联网发展治理贡献了中国方案。以习近平总书记“四项原则”“五点主张”为指引，中国发布《网络空间国际合作战略》，全面加强网络空间国际交流与合作，深度参与影响国际互联网治理进程，首次将数字经济列为二十国集团峰会主题，发布由我国牵头制定的《二十国集团数字经济发展与合作倡议》。加强以我为主的治理平台建设，创办世界互联网大会并连续举办三届，共有 4 600 多位中外各国政府官员、专家学者、企业家和国际机构负责人参会，成功搭建中国与世界互联互通的国际平台、国际互联网共享共治的中国平台。深化与联合国、亚太经合组织、金砖国家等框架内网络合作，积极推进互联网名称和数字地址分配机构国际化进程，积极推进中美、中俄、中欧等双边网络交流与合作，开展“一带一路”沿线国家网络合作，加快中国—东盟信息港、中国—阿拉伯网上丝绸之路宁夏枢纽工程等重大项目建设，推动中国治网主张成为国际共识，网络空间国际话语权和影响力显著提升。

8. 亿万人民共享互联网发展成果

习近平总书记指出，网信事业要发展，必须贯彻以人民为中心的发展思想，为老百姓提供用得上、用得起、用得好的信息服务，让亿万人民在共享互联网发展成果上有更多获得感。截至 2017 年 6 月，中国网民规模达 7.51 亿人，网站数量达 506 万个[10]。网络已经深入人们生活的方方面面，亿万网民在网上获得信息、交流信息。统筹实施网络扶贫五大工程，以互联网

10 数据来源：CNNIC，第 40 次《中国互联网络发展状况统计报告》。

助力精准扶贫、精准脱贫，网络扶贫大见成效。促进资源共享，大力发展网络教育，助力实现教育公平和效率提升，2016 年在线教育市场规模达 1 560 亿元。推动信息服务便民惠民，实现网上办事、网上购物、网上就医、网上缴费等功能，截至 2016 年年底，全国 400 多个城市借助微信、支付宝平台开通了在线城市服务，提供在线服务类型超 50 种，覆盖用户超过 4 亿。

二、全国各省区市互联网发展状况评估

本报告首次设立中国互联网发展指数指标体系，从基础设施建设、创新能力、数字经济发展、互联网应用、网络安全和网络管理六个方面，对全国各省区市互联网发展成效和水平进行科学分析、量化评估，以更全面、科学、客观地反映各地互联网发展现状和全国互联网发展总体态势，填补国内互联网领域综合性评估指标的空白。

1. 中国互联网发展指数是对习近平总书记网络强国战略思想的具体化、标准化和指标化

我们以习近平总书记网络强国战略思想为指导，以总书记在主持第三十六次中央政治局集体学习时提出的网络强国建设“六个加快”要求为主要依据（加快推进网络信息技术自主创新，加快数字经济对经济发展的推动，加快提高网络管理水平，加快增强网络空间安全防御能力，加快用网络信息技术推进社会治理，加快提升我国对网络空间的国际话语权和规则制定权），涵盖网络安全和信息化工作主要内容，设立中国互联网发展指数指标体系，旨在帮助各地进一步明确互联网发展的战略目标和重点，准确把握自身比较

优势、地域优势和发展优势，以更务实、更有效的举措落实属地责任，提高本地区互联网发展治理水平，推动全国网信事业始终朝着网络强国建设方向不断迈进。

2. 中国互联网发展指数指标体系设立立足国内、对标国际

中国互联网发展指数指标体系从本国互联网发展实际情况出发，充分吸收借鉴联合国、国际电信联盟、世界经济论坛等推出的全球电子政务发展指数、全球信息通信技术发展指数、全球网络安全指数、全球网络就绪指数等世界认可度较高的指标体系，同时参考《世界互联网发展报告 2017》中世界互联网发展指数指标体系的权重分配和计算方式，对标中国得分情况（41.8 分），得出全国 31 个省区市的互联网发展指数得分，为各省区市明确各自互联网发展水平的国内定位、国际比较提供可量化、可参考的依据。

3. 中国互联网发展指数指标体系旨在客观、真实、准确地反映各省（自治区、直辖市）互联网发展情况

为保证指标体系的科学性、准确性、权威性，中国互联网发展指数（见表 0-1）充分采纳各省区市提供的数据和材料，参考相关政府部门、行业、研究机构和企业发布的互联网领域统计数据，集纳各地区、各部门多样化数据来源，认真比对、仔细甄别、综合吸收，对全国各省区市互联网发展情况进行整体客观评估（2017 年评估范围暂未包括中国香港、中国澳门和中国台湾三地）。需要说明的是，鉴于该指数刚刚设立，尚带有一定探索性，未来将根据实际情况进一步完善指标体系、强化数据采集、扩大覆盖范围，更加全面、准确、科学地反映全国各地互联网发展情况。

表 0-1　中国互联网发展指标体系

指标名称	重点考量因素	权重	指标说明
基础设施建设	宽带接入端口数量和光纤用户占比、4G 移动基站数和 4G 用户占比、无线网络接入点个数及连接速率等	18%	反映各地宽带、移动、无线网络基础设施建设水平和网络普及情况
创新能力	专利申请数量、专利累计有效数、人力资源、研发资金投入等	16%	反映各地产业创新水平、能力和环境，以及人才培养情况
数字经济发展	ICT 产业、电信业务收入、电子商务、互联网金融、共享经济等	16%	反映各地 ICT 产业和电子商务发展情况
互联网应用	网络社交、娱乐、金融等个人应用使用率，企业网络利用率，在线政府等	25%	反映各地个人应用、企业应用使用情况和电子政务应用状况
网络安全	网络安全工作机制、网站安全漏洞防护、钓鱼网站拦截量、网民网络安全意识等	13%	反映各地 WiFi、网站和个人信息安全情况和防护水平
网络管理	网信机构设立情况、政务微博数量、政务微信公众号、网信工作人员规模等	12%	反映各地网络管理机构、队伍建设情况和管理能力

按照上述评价标准和计算方式，我们对各项指标进行赋分，得出了全国 31 个省（自治区、直辖市）的互联网发展指数得分，见表 0-2。

表 0-2　全国各省（自治区、直辖市）互联网发展评估排名表

排名	地区	基础设施建设指数	创新能力指数	数字经济发展指数	互联网应用指数	网络安全指数	网络管理指数	总分
1	广东	5.75	7.69	7.65	16.27	3.83	3.99	45.19
2	北京	6.66	5.68	5.51	18.93	3.57	4.35	44.70
3	浙江	7.34	6.73	6.46	14.60	2.93	4.31	42.36
4	江苏	7.44	7.60	6.62	13.03	3.57	3.65	41.92
5	上海	6.47	4.18	4.50	16.81	3.05	3.93	38.94
6	福建	5.53	2.89	3.89	14.46	3.66	3.00	33.43
7	四川	5.19	2.79	3.65	12.50	4.17	3.74	32.04

续表

排名	地区	基础设施建设指数	创新能力指数	数字经济发展指数	互联网应用指数	网络安全指数	网络管理指数	总分
8	山东	5.98	4.69	3.72	11.24	2.81	3.01	31.45
9	天津	5.49	3.50	2.67	12.40	3.91	2.72	30.69
10	湖北	3.94	2.69	4.35	11.86	4.39	3.11	30.34
11	安徽	2.52	3.20	3.26	11.98	5.22	2.68	28.86
12	河南	6.49	2.94	2.82	8.95	3.27	3.73	28.20
13	陕西	3.32	2.50	2.94	12.50	3.58	3.23	28.07
14	湖南	5.39	2.50	2.50	12.43	2.70	2.55	28.06
15	辽宁	6.36	2.42	3.22	10.24	2.56	3.02	27.81
16	海南	2.89	2.01	2.46	13.98	3.63	2.66	27.62
17	重庆	5.76	2.48	3.08	10.08	3.32	2.84	27.57
18	山西	4.34	2.11	2.76	10.19	4.59	2.85	26.83
19	河北	6.55	2.31	2.34	8.47	3.62	3.02	26.31
20	贵州	4.89	2.63	2.26	9.86	3.49	2.87	25.99
21	青海	3.94	2.40	2.32	9.25	5.32	2.60	25.81
22	广西	4.61	2.18	1.97	10.09	3.70	2.92	25.48
23	新疆	3.63	2.78	2.44	8.21	4.51	3.84	25.41
24	江西	4.30	2.49	2.51	9.86	2.92	2.57	24.64
25	内蒙古	4.39	1.61	2.29	8.90	4.31	2.89	24.41
26	宁夏	5.20	1.87	2.64	7.49	4.48	2.69	24.37
27	黑龙江	3.94	2.18	1.99	9.27	2.93	2.81	23.12
28	云南	1.90	2.12	2.60	10.56	4.33	1.24	22.74
29	西藏	1.97	0.77	1.69	7.46	6.99	3.49	22.37
30	吉林	4.73	1.85	1.83	8.35	2.76	2.77	22.30
31	甘肃	3.22	2.35	2.39	7.96	4.45	1.91	22.28

总体来看，从党的十八大以来，全国各省（自治区、直辖市）深入贯彻落实习近平总书记网络强国战略思想，认真按照中央统一部署要求，切实加强网上内容建设管理、网络安全保障、信息化发展等各项工作，结合本地实

际，创新发展思路，进行了一系列具有各地特色的新创造新实践，取得了显著成效。

1）在基础设施方面

各地通过加大资金投入、政策支持，探索政企合作等模式，信息基础设施普及提升，网民规模持续扩大，城乡信息基础设施差异正在缩小。总体上，信息基础设施发展水平与地方经济社会发展整体水平和投资力度呈正相关，部分地区赶超势头明显。其中，广东、四川、福建等地在光纤入户、4G 网络方面走在全国前列，贵州、湖北、山西等地积极建设数据中心，打造新型基础设施，为互联网发展奠定坚实的设施基础。

2）在创新能力方面

全国各地开始将网络信息技术创新作为推动地方创新发展的关键抓手，北京、上海等地利用人才技术优势，率先部署发展人工智能技术研究，在人工智能芯片等方面取得重大突破。内蒙古、陕西等地加大信息领域创新项目支持力度，鼓励专利申请与应用，支持科研院所、互联网企业等建立核心技术研发中心、产业联盟等，推进研发设计模式变革。贵州、山东、吉林等地支持高端信息产业发展，以领先技术支撑通信、芯片、软件等产业发展。

3）在数字经济发展方面

全国各地普遍将电子商务、共享经济作为发展重点，电子商务产业规模均呈上升趋势，地区差异相对较小，领先地区重在创新模式、培育特色、扩大规模、走向国际。浙江探索建立信息经济发展先行区，建成国际电子商务中心、全国智慧物流中心、全国数字内容产业中心等“七大中心”，数字经济规模全国领先。重庆开展省级电子商务示范城市创建工作。河南重点培育本土电商平台，在细分行业实现全国领先。

4）*在互联网应用方面*

各地对以互联网为代表的新兴技术使用度日益提升，呈现出个人应用多元化、企业应用深度化、政府应用高效化的特点。北京、上海、湖南、黑龙江等地着力打造地方新媒体，积极探索“互联网+养老”、在线教育等新模式，今日头条、澎湃、映客等客户端使用率持续攀升。浙江、广东、陕西等地企业积极利用大数据、工业云平台和工业软件等，实现智能化生产、网络化管理。内蒙古、宁夏、河南等通过建设使用草原大数据系统、农业智能联网系统等，推动农牧业生产精准化。广东建立政府网站考评机制，广西、海南等地重点建设政府信息共享平台、提供一站式在线政务服务，电子政务发展成效明显。

5）*在网络安全方面*

制定实施网络安全责任制后，特别是经过关键信息基础设施大检查、网络安全宣传周活动，各地网络安全意识和保障力度有了明显提升，但部分地区仍存在机构保障不足、人才支撑不足等问题。陕西等地加大网络安全监测力度，有效发现拦截网络攻击。北京、广东、四川等地网络安全产业发展迅猛。江苏、湖北等地积极开设网络安全学院，加大网络安全人才培养力度，具备较大发展潜力。

6）*在网络管理方面*

各地按照中央统一要求积极推动省、市两级网信管理机构建设，创新互联网管理模式，加强队伍培养，把属地管理责任落到实处。北京、上海、广东、青海等地构建网站分级分类管理体系，督促网站落实主体责任，提高管理针对性。江西把网信机构建设延伸到县一级，把网信工作情况纳入地方考核指标体系。西藏率先建立健全区地县三级网信机构，建设形成一体化的网络舆情监测、引导、治理平台体系，管网治网能力显著提升。新疆重视机构建设、干部队伍建设，违法违规信息治理有力，打击网络违法范围、侵权盗版等专项行动成效明显。宁夏、重庆、青海等立足本地互联网发展治理需要，出台了一批移动互联网管理、直播服务等方面的法规文件。

三、中国互联网发展的主要经验

中国互联网坚持一手抓发展，一手抓治理，经过 23 年波澜壮阔的发展，取得了历史性成就，发生了历史性变化，探索出了一条中国特色社会主义治网之道，为全球互联网发展治理创造了中国经验，贡献了中国智慧。总结中国互联网发展经验，主要有以下六点：

1. 坚持党对网信工作的领导

党的领导是中国特色社会主义制度的最大优势，也是中国互联网快速发展的最大优势。从 1994 年全功能接入互联网以来，中国互联网经历了从无到有、从小到大、从弱到强的快速发展过程。这一历史进程中，中国共产党不仅是中国进入互联网的决策者，更是中国互联网建设、运用、管理全过程的规划者、组织者、推动者和领航者。党的十八大以来，以习近平同志为核心的党中央高度重视网络安全和信息化工作，把完善互联网管理领导体制作为十八届三中全会确立的60项改革任务之一，成立中央网络安全和信息化领导小组，习近平总书记任组长，李克强、刘云山同志任副组长。习近平总书记主持召开中央网络安全和信息化领导小组会议、网络安全和信息化工作座谈会、中央政治局第36次集体学习等一系列重要会议，提出一系列事关网络安全和信息化发展的新理念新思想新战略，擘画中国互联网发展宏伟蓝图，为各地各部门做好网络安全和信息化工作指明了方向，提供了根本遵循。

2. 坚持以人民为中心的发展思想

互联网越来越成为人们学习、工作、生活的新空间，越来越成为人们获取公共服务的新平台。23年来，中国一直将惠及13亿多中国人民作为互联网发展的根本宗旨，让亿万民众共享互联网发展成果。通过发展互联网，加快信息化服务普及，降低应用成本，开展网络扶贫，积极适应人民期待和要求；加强网上内容建设和管理，依法打击网络违法犯罪活动，为广大网民特别是青少年营造清朗网络空间；通过网络走群众路线，运用网络更好凝聚民意、汇聚民力、开展工作，充分调动人民群众积极性主动性，共建共享互联网发展成果。

3. 坚持把创新作为互联网发展的第一动力

网信事业代表着新的生产力、新的发展方向，必须在践行新发展理念上先行一步。中国互联网贯彻落实创新、协调、绿色、开放、共享五大发展理念，始终把创新摆在首要位置，勇于创新、敢于变革，利用互联网特点和优势，加快推进体制机制创新、理念创新、技术创新、文化创新、应用创新，释放新需求，创造新供给，培育新动力，优化提升劳动力、资本、土地、技术、管理等生产要素配置效率，培育和吸引一大批互联网企业家、优秀人才、领军人才，推动新技术、新产业、新业态蓬勃发展，使互联网成为创新最活跃、发展最快速的新兴领域。

4．坚持开放与自主并举

正确处理开放与自主的关系，是中国改革开放 30 多年来的一条基本经验，也是中国互联网快速发展的重要法宝。一方面，核心技术是国之重器，中国立足自主创新、自立自强，强强联手、协同攻关，集中力量开展基础技术、通用技术，非对称技术、“杀手锏”技术，前沿技术、颠覆性技术等核心技术研究，在手机芯片、超级计算机、服务器、搜索技术等方面研发取得重大进展。另一方面，中国坚持打开国门搞建设，以开放心态引进、吸收全球网络信息先进技术与成果，有力推动中国互联网紧跟世界步伐、发展潮流，在许多领域实现了并跑甚至领跑。

5．坚持管理与服务并行

中国互联网企业由小到大、由弱变强，在稳增长、促就业、惠民生等方面发挥了重要作用。让互联网企业持续健康发展，既是企业家奋斗的目标，也是国家发展的需要。中国高度重视发挥企业在互联网发展中的主体作用，坚持鼓励支持和规范发展并行、坚持政策引导与依法管理并举，一方面，加强互联网法治建设，加快建立完善互联网领域法律法规体系，加强依法管网治网，督促企业落实主体责任。另一方面，出台了一系列支持互联网企业发展的政策措施，加快推进审批制度、融资制度、专利制度等改革，破除体制机制障碍，建立统一市场，减轻企业负担，为企业发展营造良好环境。企业坚持经济效益和社会效益并重，自觉接受政府监管，主动承担社会责任、道德责任，积极回报社会、服务人民，形成政企协作协调、共推互联网发展的合力。

6. 坚持安全与发展并重

网络安全和信息化是一体之两翼、驱动之双轮。中国始终把处理好安全和发展的关系，作为统筹推进网络安全和信息化工作的重中之重，坚持统一谋划、统一部署、统一推进、统一实施，以安全保发展、以发展促安全。树立正确的网络安全观，在信息化发展中时刻紧绷网络安全这根弦，落实网络安全责任制，开展关键信息基础设施大检查，加强自主创新、安全审查、开放合作等举措，以核心技术突破补齐安全短板，以关键信息基础设施保护加固安全屏障，切实维护整体的、动态的、开放的、相对的、共同的网络安全，为互联网快速发展提供有力的安全保障。

四、未来发展趋势与政策建议

2017 年 10 月 18 日至 24 日，中国共产党第十九次全国代表大会召开，习近平总书记代表党中央作了政治报告，科学分析当今世界和当代中国发展大势，进一步明确“两个一百年”奋斗目标，提出新“两步走”战略，描绘全面建设社会主义现代化国家总蓝图，对党和国家事业发展做出全局性、战略性、前瞻性规划部署，开启了中国特色社会主义新时代。报告对推动互联网、大数据、人工智能、共享经济发展，建设网络强国、数字中国、智慧社会等作出重大战略部署，为中国互联网发展提供了根本遵循与行动指南。未来三十年，是中国互联网大有可为的重大战略机遇期。站在新的历史起点上，中国互联网必须坚持以习近平新时代中国特色社会主义思想为指导，全面贯彻落实党的十九大精神，认真贯彻以人民为中心的发展思想，加快实施网络强国战略，大力推进数字中国建设，努力实现中国互联网更全面、更平衡、更充分、更高质量的可持续发展，让互联网更好地造福国家和人民，为推进中国特色社会主义现代化作出新的更大贡献。

1. 互联网发展将从加速普及转向质量提升，应以“数字中国”建设为统领铸就经济社会发展新设施新平台

中国互联网经过 23 年的高速发展，正进入量变到质变的关键期。当前阶段，互联网普及省域差异逐渐缩小，用户规模增幅持续收窄，社会资本开始趋于理性，依靠既有的设施红利、人口红利、资本红利换取高速增长的可持续性有限。未来，应高标准、高质量、高水平规划建设“数字中国”，充分发挥互联网对深化供给侧结构性改革的驱动引领作用，全面升级网络基础设施，加速推进新一代信息通信技术商用部署，加快建设下一代互联网，全面布局 IPv6 建设，推动网络基础设施向高速、融合、智能、开放的目标演进，构筑全面提升经济发展质量和效益的基础设施平台、数据平台、资源平台，促进一/二/三产业发展迈向全球价值链中高端。

2. 人工智能将成为网络信息技术发展的新“蓝海”，应以关键性、基础性软硬件研发为重点抢占制高点

在全球信息技术发展的新一轮创新活跃期，云计算、大数据、物联网、人工智能等新一代信息技术迭代加速，特别是人工智能作为一门融合计算机科学、统计学、脑神经科学和社会科学的前沿综合性学科，将引领下一波世界科技革命新浪潮，引发各行各业的深度技术变革，带动技术、数据、产业等方面的爆发式飞跃。技术方面，中国高性能计算能力快速提升、深度学习算法不断演进、移动互联网提速加快，这些技术综合能力的飞跃为人工智能技术突破创造了可能，应集中力量研发突破关系人工智能技术发展的基础算法与芯片/处理器技术、智能操作系统、计算机视觉与图像技术、高速计算与大规模数据处理技术、智能机器人等重要关键技术，占领新一轮信息技术发展制高点。产业方面，中国海量用户数据将成为人工智能发展的绝对优势，人工智能的数据计算、处理和应用将为社会创造更多高价值数据，应将投资重心从“互联网+”向“人工智能+”转移，掀起新一轮创新创业热潮，推出一批满足多应用场景和智能服务需求的现象级人工智能产品，带动各行业领域产业组织方式、服务模式、商业模式的创新变革。

3．数字经济发展方兴未艾，应抓住新一轮经济全球化趋势进一步拓宽发展空间

中国数字经济发展速度和发展水平已处于全球领先位置，且具备持续增长潜力。未来，随着互联网、大数据、人工智能和实体经济的深度融合，以数据驱动为特征的新一轮经济全球化将加速推进，数据贸易全球化、电子商务全球化、互联网金融全球化、在线服务全球化不可避免，跨境平台交易量、需求量、用户量有望实现爆发式增长，更多中小企业甚至个人将分享经济全球化红利，平台经济体将取代传统跨国企业，成为最具影响力的国际贸易平台。中国应抓住机遇加快建设融通中外、惠及全球的新型开放型数字经济体制（机制），一方面，加快推动智慧农业、智能制造、在线服务、互联网金融等数字产业发展，在中高端消费、共享经济、现代供应链、人力资本服务等领域培育新增长点、形成新动能；另一方面，抓紧研究制定跨境数据贸易、电子商务、服务贸易、互联网金融等方面的国际规则与标准，推动成立全球数字经济发展联盟，引导中国互联网企业积极参与“一带一路”、自由贸易港、多边双边自贸区等国际经济带建设，进一步拓展数字经济发展空间，提升全球数字经济发展水平。

4．内容付费将成为互联网媒体盈利增长新热点，应积极推动中华文化创造性转化、创新性发展

随着网络空间版权意识的增强，数字内容消费习惯的形成，网络付费订

阅模式将不断趋热，涉及新闻、视频、文学、社交等网络媒体各个领域，使得原创内容、优质内容价值变现更便利、更高效，将催生更大的数字内容生产、服务、消费市场。“内容为王”重要性日益凸显，传统媒体可利用内容生产优势，寻找转型突破的新契机，扩大内容影响力和收益率。新兴媒体可通过优化用户体验，深耕专业细分领域，抢占独家版权资源，不断提升内容“溢价”。中华文化源自五千多年文明历史，融合了革命文化、社会主义先进文化，正越来越受到国际社会和世界人民的认同与喜爱，应积极打通境内外各类优秀网络平台，积极利用人工智能、付费订阅等新型网络技术和商业模式，推动中华文化创造性转化、创新性发展，不断提升国际传播能力，讲好中国故事，展现真实、立体、全面的中国，提高国家文化软实力。

5．智慧社会建设事关党和国家领导管理体制机制改革成效，应以推动数据资源开放利用促进国家治理体系和治理能力现代化

随着移动互联网深入发展、智能手机日益普及、电信普遍服务全面入村入户，智慧社会建设将加快推进，在线化服务、网络化生活、平台化治理成为经济社会发展新模式，将深刻影响经济、政治、文化、社会、生态文明建设进程。未来，电子政务建设模式将更加集约，全国一体化的电子政务平台成为建设主流，推动数据资源开放共享，打通政府部门之间、网络平台之间、政府与网络平台之间的数据壁垒，实现跨层级、跨地区、跨部门、跨系统的政府管理与协同治理变得十分必要。政务信息资源开发利用将持续深化，从静态存储、简单应用向动态管理、深度剖析转变，应通过大数据、人工智能等技术，提前发现社会潜在的风险隐患、及时修补监管漏洞，大力提高社会治理效能。智慧社会建设将从个别试点向全面铺开转变，推动社会治理和公共服务更加个性化、智能化、精准化，更好满足公众多元化需求和对美好生活的向往。

附：英文版总论

Overview

Looking at the history of world civilizations, humanity has progressed along with agricultural, industrial and information revolutions. Each of these industrial and technological revolutions has had great and profound impact on our way of production and life. Today, information technologies represented by the Internet are experiencing rapid changes with each passing day. They have brought about new ways of social production, created new space for people's life, opened new horizons of state governance, and enhanced our ability to understand and shape the world. China is going through a historic process of rapid application of information technologies. The country attaches great importance to Internet development. Since it was connected to World Wide Web 23 years ago, we have, in keeping with the principles of proactive utilization, rational development, law-based regulation and assurance of security, strengthened IT infrastructure, developed digital economy and made life better for our people through IT application.

Since the 18th National Congress of the Communist Party of China (CPC), the CPC Central Leadership, with Xi Jinping as the core, holding the right banner, has been making the right strategies, conquering one difficult after another, forging ahead and launching reforms and innovations. They try to make everything better by considering the overall domestic and international situation. They have been planning and promoting the general layout of "Five in One" (*economic, political, cultural, social, and ecological progress*), coordinating and promoting the strategic layout of "Four Comprehensives" strategy (*to comprehensively deepen reform, comprehensively advance rule of law, and comprehensively strictly govern the Party so as to provide strong impetus and guarantee for comprehensively building a moderately prosperous society*), carrying out new development ideas, further reforming difficult areas, and promoting ethical party and government construction and combating corruption. Thus many problems have been solved that remained unsolved before, and many desirable things have been done that remained undonc, leading to historic changes in the CPC and state governance, and to the establishment of the guiding position of Xi Jinping Thought on Socialism with Chinese Characteristics for a New Era. The CPC Central Leadership stresses cyber security and information security, listing the perfection of the cyberspace leadership mechanism as one of the sixty reform tasks defined at the Third Plenum of the 18th CPC Central Committee. Thus the Central Leading Group for Cyber Affairs was formed, who integrates and coordinates cyber security and informatization issues in politics, economy, culture, society and military affairs. CPC General Secretary Xi Jinping, a visionary man, has put forward a series of new ideas, thoughts and strategies based on China's practice of Internet governance, systematically illustrating the important theoretical and practical issues concerning cyber security and informatization and thus forming rich, profound, scientific and systematic strategic thoughts on developing and strengthening China through Internet development. On the 19th CPC National Congress, the "Two Step Development Strategy" was put forward, ushering in a new era of building socialism with Chinese characteristics. The Congress has made strategic deployments on promoting development of Internet, big data, AI, and sharing economy so as to turn China into an influential country in cyberspace and a leading one in digital development and smart society, which will guide China's Internet development in years to come.

I. Achievements of China's Internet Development

Guided by Xi Jinping Thought on Socialism with Chinese Characteristics for a New Era, and his strategic thought on developing and strengthening China through Internet development in particular, the top-level design and general framework for the country's Internet development and governance have been basically finished, positive energy and main theme spread has become the trend in the cyberspace, which is becoming cleaner day by day. China's cyber security has been consolidated. In the country, informatization plays an outstanding role in driving and leading the economic development. Meanwhile, China's power of discourse and influence concerning cyberspace in the world have increased, and the Chinese people have got more sense of gain in sharing the results of Internet development. All in all, we have achieved a lot in Internet development.

1 . Network infrastructure has become a new type of public infrastructure that promotes economic and social development

General Secretary Xi Jinping points out that we should strengthen network infrastructure construction and in-depth information resource integration, and open up the "main artery" of information for economic and social development. Aware of the trend of the global Internet development, China is now implementing the "Broadband China" Strategy and launching the action of "Facilitating Faster and More Affordable Internet Connection". We are constantly promoting network infrastructure construction and upgrading, and improving the capacity of broadband access, Internet service quality and application level. We are the leader in the world in broadband network scale and construction quality. By June 2017, the bandwidth of China's backbone network had exceeded 100Tbps, and that of international export had reached 7,974.779Gbps; the number of the Internet broadband access ports had reached 739 million; that of fixed broadband users, 322 million; and that of mobile communication, 1.36 billion, including 890 million 4G users[11]. Network infrastructure in rural areas has witnessed leapfrog development, general service in telecommunication has improved greatly, and all townships and 93.5 percent of administrative villages have access to broadband. Application infrastructure is developing fast, the node number of Content Distribution Network (CDN) has reached over 2,400, and peak broadband reserve has exceeded 10TB. The quantity of key Internet resource has seen great growth, with the number of IPv4

11 Data source: Economic Performance of Telecommunications Jan. to Jun., 2017, Ministry of Industry and Information Technology of the People's Republic of China (MIIT); The 40th China Statistical Report on Internet Development, China Internet Network Information Center (CNNIC).

addresses amounting to 338 million, that of IPv6 to 21,283 (in terms of blocks)/ 32 (in terms of addresses), both ranking second in the world. The number of registered domain names (.cn) is 20.61 million[12], ranking first in the world.

2. Breakthroughs have been made in Internet information technology

General Secretary Xi Jinping points out that China should speed up its self-dependent innovation of core technologies in information area and thus to narrow the gap between China and developed countries and even to become the leader of the world in some IT areas. According to Xi, by implementing the innovation-driven development strategy, China should launch a series of key projects with long-term benefits to the overall development of the country, carry out with unified strength researches into key generic technologies of the Internet and into cutting-edge technologies, trying to make breakthroughs in high-performance computing, quantum communication, artificial intelligence, cloud computing, big data, kernel processor chip, operating system and 5G technology. By the end of 2016, the country had had 227,365 valid invention patents in computer, communication and other electronic equipment manufacturing, including 118,725 in the single year of 2016, putting China on the top of the world in terms of international patent registration four times in recent five years. Mozi Satellite, China's first quantum science experimental satellite, has been successfully launched into the sky; Beijing-Shanghai Trunk, the world's first quantum confidential communication network, has been opened, covering over 2,000 kilometers; and Beidou Navigation System (BDS), a global satellite navigation system researched and developed by China itself, has been applied in

12 Data source: The 40th China Statistical Report on Internet Development, CNNIC.

a large scale, with two Beidou No. 3 satellites launched at the first time, leading the country into a new era of globally networking. China has taken active part in the formulation of international telecommunication technology standards, playing a leading role in completing a number of IETF standards, making key breakthroughs in IPv4/IPv6 interconnection, cyber security and routing protocols, and producing increasing influence on the next-generation core technologies of the Internet.

3. New economy represented by digital economy is thriving

According to General Secretary Xi Jinping, cyber and information technology sector represents the new productivity and new development direction, so it is expected to take one step ahead in practicing the new development philosophy. China gives full play to informatization's role in driving and leading the development of economy and society and develops its digital economy, having made remarkable achievements in e-commerce, information service, network terminal equipment, Internet finance, sharing economy and Internet+ integration. In 2016, the country's digital economic gross scale reached RMB 22.58 trillion yuan, leaping to the second place of the world, and accounting for 30.3 percent of national GDP[13]. Digital economy has become the new engine and new highlight of China's economic growth. E-commerce, the most active sphere in China's digital economy today, involved 467 million online shoppers in 2016,

13 Data source: White Paper of China's Digital Economy Development 2017, China Academy of Information and Communications Technology, (CAICT).

with its transaction volume amounting to RMB 26.1 trillion yuan[14] and online retail sales volume to RMB 5.2 trillion yuan, ranking first in the world. Sharing economy is booming, with the market turnover amounting to RMB 3.45 trillion yuan[15] in 2016. China's Internet finance is also developing rapidly, with the number of its electronic payment services amounting to 139.56 billion in 2016, and thus a service value of RMB 2,494.5 trillion yuan[16]. The country encourages, supports and regulates the development of digital economy through political guiding and administration under laws to create a favourable market environment for information network business, which has led to the increase of international influence of a number of companies such as Alibaba, Tencent, Baidu, and Huawei, and the emergence of many new Internet companies with innovation energy. As of October 2017, 730 Chinese companies in this area have gone public at home or abroad, with a total market value exceeding RMB 17.5 trillion yuan. Among the world's top 20 Internet companies in terms of market value, there are eight from China.

4. The development of e-government has helped to innovate social governance

General Secretary Xi Jinping stresses that we should speed up our e-government and encourage the governments of different levels to break the information barrier and improve their service efficiency so that the common

14 Data source: Report on China's E-commerce 2016, Ministry of Commerce of the People's Republic of China (MOFCOM).

15 Data source: Report on China's Sharing Economy Development 2017, Research Branch of Sharing Economy, State Information Center of China (SIC) and Sharing Economy Working Committee, Internet Society of China (ISC).

16 Data source: Overview of Payment System Operation, The People's Bank of China (PBOC).

people will made less effort to get more information even from afar and so that we can make things easier, faster and simpler. China has listed "Internet +e-government" into the 12 priorities under the *National Informatization Planning for the Thirteenth-five-year Plan Period*, having formulated policies such as *Guidelines on Promoting the Coordinated Development of E-government* and *Guidelines on Speeding up the Internet + E-government*. All the local governments are expected, in accordance with the planning made by the Central Government, to keep deepening the application of e-government, establish the mechanism of governmental interlink, information sharing and business interconnection, promote the construction of online service platforms for public service integration, and provide the full-flow, one-stop online service across sectors and areas. By the end of 2016, out of 31 provinces, autonomous regions and municipalities directly under the Central Government, and Xinjiang Production and Construction Corps, 27 had established the provincial e-government service platform[17]. New media of e-government are witnessing rapid development. By June 2017, there had been 170,000 e-government weibo accounts verified by SinaWeibo Open Platform, and 514, 000 e-government WeChat public accounts. They are new means of enhancing information openness, interaction between the governments and the people, and supervision from the people, guaranteeing the citizens' rights to know, to participate, to express their opinions and to supervise the governments.

17 Data source: Blue Book of E-government: Report on China's E-government Development 2017, Social Sciences Academic Press of China.

5. The capacity of cyber security safeguarding keeps improving

General Secretary Xi Jinping points out that without cyber security there would be no national security and without informatization there would be no modernization. According to Xi, security ensures development, and development is what security is aimed at. China has formulated a series of laws, regulations, strategies and plans including the *Cyber Security Law of the People's Republic of China* and the *National Cyber Security Strategy*, established and strengthened the cyber security accountability, cyber security review mechanism, and cyber security monitoring, warning and response mechanism, amended the *National Emergency Response Plan for Internet Security Incidents,* and enhanced cyber security awareness and capacity of incident analyzing and tracing as well as of quick recovery after cyber attacks. The country has reinforced the security protection of key network infrastructure, and has, for the first time, launched nationwide inspection on the key network infrastructure in the key areas like finance, energy, electric power, communication and transportation to promote improvement and construction. It is speeding up standardization of cyber security. By the end of 2016, the National Technical Committee for Information Security Standardization (NTCISS) had issued 195 information safety standards[18]. The cyber security industry scale keeps increasing, with over 1,000 such enterprises and RMB 135.5 billion yuan[19] of value in 2017. The first-level disciplines of cyber security have been established in universities. Seven demonstration projects in cyber security colleges have been launched and cyber security schools

---

18 Data source: CCID.

19 Data source: Blue Book of China's Cyber Security Development 2016-2017, China Academy of Electronics and Information Industry Development.

have been established in 21 universities. Insisting that the cyber security should serve and rely on the people, we attach importance to cyber security publicity and education, having held four "National Cyber Security Publicity Weeks", involving over two billion participants, and thus the cyber security awareness and techniques have been dramatically improved nationwide.

6. The cyberspace is becoming clearer day by day

General Secretary Xi has pointed out that the cyberspace is the common home of humankind. Making it better, cleaner and safer is the common responsibility of the international community, according with the interests of the people. Positive energy spread in cyber content construction and administration is the general requirement, and the controllability is the absolute principle. We are enhancing positive publicity, innovating the dissemination means, managing the timeliness, appropriateness and efficiency of public opinion in cyberspace, improving the transmissibility, leadership, influence, and credibility, and launching a large number of phenomenal news broadcasts and thematic publicity products conforming to the online dissemination features and the interest of users. All of these measures have helped to increase the rate of arrival, reading and thumbing up. We also pay attention to the cultivation and practice of the core socialist values. For example, we have launched the online dissemination of excellent Chinese culture and cultivate healthy online culture, so that the core socialist values can be adopted to increase cohesion of society. We have adopted legal, administrative and technical means to strengthen the cyberspace administration, improve the network communication law enforcement system and mechanism, promote Internet governance, opening and visiting in accordance with laws, and enhance the accountability of competent authorities for the local cyberspace administration and the responsibility of website entities, and advocate administration in a scientific and regulated way. Full play has been given to the roles of social organizations, enterprises and specialists, and the mode of network environment management involving different participants has been formed. The mechanism for linked tackling of illegal information and websites, and complaints from users has been improved. In 2016, 40 million complaints were tackled, and more than 92 percent of harmful information was disposed of. So far, twenty specific projects like "cleaning" and "seedling-protection" have been launched to dispose of harmful information, which has helped to make cyberspace cleaner.

7. China's power of discourse and influence in cyberspace keeps increasing

General Secretary Xi Jinping has illustrated the concept of the international cyberspace administration on many an occasion. He has proposed "four principles" for promoting the global Internet governance system reform and "five proposals" for constructing a community of shared future in cyberspace, which has contributed to global Internet development and governance. Guided by Xi's "four principles" and "five proposals", China has released *Strategy of International Cooperation in Cyberspace*. It enhances its communication and cooperation with other countries, and participates in and produces influence on the progressing of the international Internet governance. It has listed "digital economy" as the theme of G20 Summit for the first time in the history of G20 Summit, and released the *G20 Digital Economy Development and Cooperation Initiative*, in the formulation of which China was the coordinator. We have enhanced the almost totally self-reliant construction of the Internet governance platforms and opened and hosted three times the World Internet Conference (Wuzhen Summit), with 4,600 participants, including officials, specialists, entrepreneurs, and leaders from different countries and international organizations. The Wuzhen Summit has become the international platform for interconnectivity between China and other countries and the Chinese platform for Internet sharing and shared governance. We have deepened the cyberspace cooperation in the framework of the UN, APEC, and BRICS, promoted the internationalization of Internet name and digital address distribution mechanisms, as well as bilateral communication and cooperation between China and the US, Russia and Europe. Cooperation has been launched among countries along the Belt & Road, the construction of key projects like China-ASEAN Information Harbor, China-Arab States Online Silk Road Ningxia Pivot has been accelerated, and China's views on Internet governance have been made into common consensus. The country's power of discourse and influence in cyberspace has increased dramatically.

8 . The Chinese people are sharing the benefits of Internet development

General Secretary Xi Jinping points out that the development of the Internet development must be oriented to benefit the people, who can afford to use the Internet and then have the chance to use it and use it well, thus having the sense of gain in enjoying the benefits of Internet development. By June 2017, the number of China's Internet users had reached 751 million and that of websites in the country, 506 million[20]. The Internet has become an indispensable part in our

20 Data source: The 40th China Statistical Report on Internet Development, CNNIC.

daily life, with hundreds of millions of Internet users obtaining and exchanging information on it. We have implemented as a whole five projects of poverty alleviation through the Internet, which has seen great success thanks to the targeted poverty alleviation and elimination. We are promoting resource sharing and developing education through the Internet to realize education equality and improve the efficiency. In 2016, the online education market scale was RMB156 billion yuan. We are making information service convenient and beneficial to the people, who then can go through official procedures, do shopping, see the doctor and pay on the Internet. By the end of 2016, over 400 cities had opened up online city service through WeChat and Alipay, covering 50 types of online service and over 400 million users.

II. Assessment of the Internet Development in all Provinces, Autonomous Regions and Municipalities Directly Under the Central Government

In this report, we set up China's Internet Development Index System covering infrastructure construction, innovation capacity, digital economic development, Internet application, cyber security and cyberspace regulation. We have made an objective, authentic and accurate assessment of the effect and level of the Internet development in all provinces, autonomous regions and municipalities directly under the Central Government. Scientific analysis and quantitative assessment have been adopted as the method. It is a comprehensive assessment index system in the Internet area, the first of its kind in China.

1. China's Internet Development Index is the specification, standardization and indexation of General Secretary Xi Jinping's strategic thought on developing and strengthening the country through the Internet

Guided by Xi's thought, and based on the "six accelerations" (namely, to accelerate self-dependent innovation in network information technologies, to accelerate digital economy's drive for economic development, to accelerate the improvement of cyberspace administration, to accelerate cyber security protection, to accelerate the promotion of social governance through network information technologies, and to accelerate the enhancement of China's power of discourse and rule-making right in the world in terms of cyberspace) put forward by President Xi when he presided over the 36th Collective Study Session of the Political Bureau of the CPC Central Committee, our assessment covers the major content of cyber security and informatization. The Index System will make local governments more aware of their strategic goals and focus of Internet development, as well as their comparative, geographical and development advantages, so that they can take more practical and efficient measures to fulfill the local responsibilities, improve their local Internet development and governance and promote the development of the national cyberspace administration and information sector in the direction defined by Xi Jinping.

2. China's Internet Development Index focuses on the domestic situation through a comparison with the international situation

Based on the domestic situation, by referring to the E-Government Development Index (EGDI), ICT Development Index, Global Cyberspace Security Index, and Networked Readiness Index (NRI) released by the UN, International Telecommunication Union (ITU) and World Economic Forum

(WEF), and referring to the weight distribution and calculation method contained in the World Internet Development Report 2017, we have got the scores of the Internet Development Index of 31 provinces, autonomous regions and municipalities in accordance with China's score (41.8), providing measurable and referable basis for all provinces, autonomous regions and municipalities.

3. China's Internet Development Index is an objective, authentic and accurate assessment of the effect and level of the Internet development in all provinces, autonomous regions and municipalities directly under the Central Government

To ensure the scientificity, accuracy and authority of the Index(see Tab.0-1), we have adopted all the data and materials provided by the local authorities and referred to statistics and materials provided by local departments, industries, research institutes and enterprises. But, due to the limited data collection, our assessment in 2017 does not involve Hong Kong, Macao and Taiwan. Besides, in view that the system has just been established, it is to be tested. We will adjust and improve the index system, enhance data collection and expand the coverage, so that the assessment will reflect the reality more completely, accurately and scientifically.

Table 0-1 China's Internet Development Index System

Indicator	Key assessment factors	Weight	Specification
Infrastructure construction	Number of broadband access ports and percentage of optical fiber users, number of 4G mobile base stations, number of wireless network accesses and access speed	18%	Local infrastructure construction level and Internet coverage of broadband, mobile and wireless network
Innovation capacity	Number of patents registered, accumulated patent efficiency, human resources, and investment in R &D.	16%	Level , capacity and environment of local industrial innovation and local talents cultivation

续表

Indicator	Key assessment factors	Weight	Specification
Development of digital economy	ICT industry, income of telecommunication, e-commerce, Internet finance, and sharing economy	16%	Development of local ICT industry and e-commerce
Internet Application	Rate of individuals' use of the Internet for socializing, entertainment and finance, rate of corporations' use of the Internet, and online governments	25%	Local individuals' use and corporations' use of the Internet, and e-government
Cyber security	Cyber security working mechanism, cyber security holes protection, phishing site intercept, and Internet users' cyberspace security awareness	13%	Security and protection of local WiFi, websites, and personal information
Cyber regulation	Establishment of cyber organizations, number of governmental Weibo and WeChat accounts, and the number of staff engaged in Cyber affairs	12%	Capability of constructing and managing local cyber regulation organizations and personnel

Following the above weight distribution and calculation method, we have got the scores of the Internet Development Index of 31 provinces, autonomous regions and municipalities of China(see Tab.0-2).

Tabel 0-2 Ranking of the Provinces, Autonomous Regions and Municipalities in the Assessment of the Internet Development

Ranking	Region	Index of infrastructure construction	Index of innovation capacity	Index of development of digital economy	Index of Internet application	Index of cyber security	Index of cyberspace regulation	Total score
1	Guangdong	5.75	7.69	7.65	16.27	3.83	3.99	45.19
2	Beijing	6.66	5.68	5.51	18.93	3.57	4.35	44.70
3	Zhejiang	7.34	6.73	6.46	14.60	2.93	4.31	42.36
4	Jiangsu	7.44	7.60	6.62	13.03	3.57	3.65	41.92
5	Shanghai	6.47	4.18	4.50	16.81	3.05	3.93	38.94
6	Fujian	5.53	2.89	3.89	14.46	3.66	3.00	33.43

续表

Ranking	Region	Index of infrastructure construction	Index of innovation capacity	Index of development of digital economy	Index of Internet application	Index of cyber security	Index of cyberspace regulation	Total score
7	Sichuan	5.19	2.79	3.65	12.50	4.17	3.74	32.04
8	Shandong	5.98	4.69	3.72	11.24	2.81	3.01	31.45
9	Tianjin	5.49	3.50	2.67	12.40	3.91	2.72	30.69
10	Hubei	3.94	2.69	4.35	11.86	4.39	3.11	30.34
11	Anhui	2.52	3.20	3.26	11.98	5.22	2.68	28.86
12	Henan	6.49	2.94	2.82	8.95	3.27	3.73	28.20
13	Shaanxi	3.32	2.50	2.94	12.50	3.58	3.23	28.07
14	Hunan	5.39	2.50	2.50	12.43	2.70	2.55	28.06
15	Liaoning	6.36	2.42	3.22	10.24	2.56	3.02	27.81
16	Hainan	2.89	2.01	2.46	13.98	3.63	2.66	27.62
17	Chongqing	5.76	2.48	3.08	10.08	3.32	2.84	27.57
18	Shanxi	4.34	2.11	2.76	10.19	4.59	2.85	26.83
19	Hebei	6.55	2.31	2.34	8.47	3.62	3.02	26.31
20	Guizhou	4.89	2.63	2.26	9.86	3.49	2.87	25.99
21	Qinghai	3.94	2.40	2.32	9.25	5.32	2.60	25.81
22	Guangxi	4.61	2.18	1.97	10.09	3.70	2.92	25.48
23	Xinjiang	3.63	2.78	2.44	8.21	4.51	3.84	25.41
24	Jiangxi	4.30	2.49	2.51	9.86	2.92	2.57	24.64
25	Inner Mongolia	4.39	1.61	2.29	8.90	4.31	2.89	24.41
26	Ningxia	5.20	1.87	2.64	7.49	4.48	2.69	24.37
27	Heilongjiang	3.94	2.18	1.99	9.27	2.93	2.81	23.12
28	Yunnan	1.90	2.12	2.60	10.56	4.33	1.24	22.74
29	Xizang (Tibet)	1.97	0.77	1.69	7.46	6.99	3.49	22.37
30	Jilin	4.73	1.85	1.83	8.35	2.76	2.77	22.30
31	Gansu	3.22	2.35	2.39	7.96	4.45	1.91	22.28

In general, since the 18th CPC National Conference, the whole nation has been carrying out General Secretary Xi Jinping's strategic thought on developing and strengthening the country through the Internet. They are, in accordance with the unified arrangement by the CPC Central Committee, enhancing the cyber content construction, cyber security guarantee, and informatization through a series of new creations and practices based on their local reality, and have achieved great success.

1）In terms of infrastructure

The coverage of information infrastructure and the number of Internet users have increased, and the gap between urban and rural areas in information infrastructure is narrowing since the local governments have increased the fund and political support. In general, there is positive correlation between the information infrastructure development and the local economic and social development and investment level, and in some regions, the information infrastructure development is even ahead of the local economic and social development and investment level. For example, Guangdong, Sichuan and Fujian take the lead in the FTTH (Fiber to The Home) and 4G network, Guizhou, Hubei and Shanxi are constructing their data center and new-type infrastructure to lay a solid infrastructure foundation for the Internet development.

2）In terms of innovation capacity

All regions participating in the assessment have begun to take network information technology innovation as the grip for the local innovative development. For example, Beijing and Shanghai, by taking the advantage of having more talents and technology, take the lead in developing artificial intelligence technology, having made breakthroughs in artificial intelligence chips; Inner Mongolia and Shaanxi have increased their support for innovation projects in the areas concerning information, encourage patent registration and application, and support research institutes and Internet enterprises in establishing core technology R&D centers and industrial alliances, and promote R&D design mode revolution; Guizhou, Shandong and Jilin support high-end information industry development, and especially support telecommunication, chip and software industry development with leading technology.

3) In terms of development of digital economy

All regions participating in the assessment take e-commerce and sharing economy as the development focus, so e-commerce is increasing and seeing little regional difference. Regions leading in e-commerce attach importance to mode innovation, cultivate unique features, increase the scale and go global. For example, Zhejiang takes the lead in digital economy scale by establishing experiment and advance zones for information economy development, and seven centers including the International E-commerce Center, National Intelligence Logistics Center, and National Digital Content Industry Center. Chongqing is trying to become one of the national pilot cities of the provincial level in terms of e-commerce, and Henan is cultivating local e-commerce platforms, trying to take the lead in the country in terms of industrial segment of e-commerce.

4) In terms of Internet application

Application dimension of new technologies represented by the Internet has been increasing, with diversified individual application, in-depth industrial application and efficient governmental application. For example, Beijing, Shanghai, Hunan and Heilongjiang are fostering their new local media and trying to establish new models like Internet+ senior citizens caring and online teaching. The client use ratio of Toutiao, The Paper and Inke keeps increasing. Enterprises in Zhejiang, Guangdong and Shaanxi have launched intelligent production and network-based management by using big data, industrial cloud platforms and industrial software. Inner Mongolia, Ningxia and Henan are promoting precise production of agriculture and animal husbandry by using the prairie big data system and agricultural intelligent networking system. Guangdong has set up the assessment mechanism for governmental websites, and Guangxi and Hainan have set up governmental information sharing platforms and provide one-stop online government service, so their e-government has seen great success.

5）In terms of cyber security

Since the cyber security accountability system was adopted, especially since the Key Information Infrastructure Inspection and China Cyber Security Week were carried out, the cyber security awareness and guarantee strength have obviously increased in different regions, but there remain problems like insufficient organization guarantee and insufficient talents support. Provinces like Shaanxi has strengthened cyber security monitoring and found and blocked cyber attacks. Beijing, Guangdong and Sichuan have seen rapid development of cyber security industry. Jiangsu and Hubei have established cyber security colleges to enhance the cultivation of talents in that field, with great potential in development.

6) In terms of cyberspace regulation

All local governments are, in accordance with the unified requirement of the Central Government, promoting the establishment of provincial and prefectural cyberspace administration organizations, innovating the cyberspace administration patterns, strengthening team-building and fulfilling local accountability. For example, Beijing, Shanghai, Guangdong and Qinghai have constructed website classification and categorization administration systems, urging websites to fulfill their accountability and increasing the pertinence of the administration. Jiangxi has extended cyberspace administration organization construction to the county level, and channeled cyberspace administration into the local assessment indicator system. Tibet became the first to set up and improve three-level cyberspace administration organizations, namely, the autonomous region level, the prefectural level and the county level, which has helped to form an integrated cyberspace public sentiment monitoring, guidance and governance platform system and thus has improved the cyberspace administration capacity. Xinjiang attaches importance to institutional construction and team-building, punishment and governance of illegal information spread. They have achieved great success in cyberspace crime punishment and punishment of piracy and patent infringement. Ningxia, Chongqing and Qinghai have, in accordance with the local needs for the Internet development and governance, launched a series of laws and regulations on mobile Internet administration and webcast service.

III. China's Experience in Internet Development

China attaches importance to both development and governance of the

Internet. After 23 years of development, the country has made historic achievements, witnessed historic changes and found a path of Internet governance with Chinese characteristics, thus contributing China's experience and wisdom to the global Internet development and governance. The country's experience in Internet development can be summarized as follows:

1. Ensuring the CPC's leadership in cyberspace affairs

The leadership of the Communication Party of China (CPC) is the greatest advantage for the socialist system with Chinese characteristics and for the rapid development of China's Internet. Since the country's full-function access to the Internet in 1994, its Internet has developed from nothing to something, from small to big and from weak to strong. During the process, the CPC is not only the decision-maker on the country's access to the Internet, but also the planner, organizer, promoter and navigator of its network construction, development and administration. Since the 18th CPC National Congress, the Central Leadership of CPC with Xi Jinping as the core has been attaching great importance to cyber security and informatization. They have taken the improvement of the cyberspace administration leadership mechanism as one of the 60 reforms defined at the Third Plenum Session of the 18th CPC National Congress. They founded the Central Leading Group for Cyberspace Affairs, of which Xi is the director and Li Keqiang and Liu Yunshan the deputy directors. General Secretary Xi Jinping has chaired a series of meetings and symposiums, such as the Meeting of the Central Leading Group for Cyberspace Affairs, the Symposium on Cyber Security and Informatization, and the 36th Collective Study Session of the Political Bureau of the CPC Central Committee, putting forward a series of new ideas, thoughts and strategies about cyber security and informatization, drawing the great blueprint for China's Internet development, and thus defining the direction and guidelines for cyber security and informatization of all sectors.

2. Committing to the people-centered development philosophy

The Internet has become the new space for people's study, work and life, and the new platform for obtaining public services. Over the past 23 years, China has been pursuing the basic principle of developing the Internet to benefit its people. By developing the Internet, the country is speeding up the coverage of information service, reducing the application cost, carrying out poverty alleviation, and meeting the expectation and need of the people. It is enhancing content construction and governance of the Internet, punishing illegal online actions and crimes, and creating clean cyberspace for Internet users, especially young people. The Party is carrying out the mass line, using the Internet to gather the public opinions, unite the people, carry out some work and mobilize the initiative and enthusiasm of the people, who contribute to and enjoy the Internet development results.

3. Taking innovation as the first driver for Internet development

Cyberspace administration represents new productivity and new development direction, so it must be ahead in new idea practice. In developing its Internet, China is carrying out Five Development Concepts, namely, innovation, coordination, greenness, openness and sharing, with innovation always in the very first place. The country is brave enough to innovate and reform. It is speeding up the innovation in system and mechanism, ideas, technology, culture and application by making use of the features and advantages of the Internet. It releases new demand, creates new supply, cultivates new drive, optimizes the allocative efficiency of labour, capital, land, technology and management. It also cultivates and attracts a large group of Internet entrepreneurs, talents, leading figures in this field, and enhances the thriving of new technologies, new industries, and new ecology, so that the Internet has become the rising field with the most active innovation and the most rapid development.

4. Pursuing both opening up and self-reliance

To tackle the relationship between opening up and self-reliance is the basic experience of China over the past 30 years and the solid foundation for the rapid Internet development of the country. As is known, core technology is a nation's treasure, so China, with self-dependent innovation, self-reliance and self-improvement, would like to cooperate with others in developing fundamental technologies and general technologies, asymmetric techniques and trump-card techniques, core technologies like cutting-edge technology and game-changing technology, and thus we can make breakthroughs in R&D of chips for cell phones, super computers, servers and searching technologies. Besides, China keeps constructing everything in opening up and introduces and absorbs advanced technologies and fruits from global network information areas, and makes its Internet development keep up with the world's pace and trend, so that the country has become the leader in many areas of the Internet.

5. Pursuing paralleled development of regulation and service

China's Internet enterprises have been developing from small to big and from weak to strong, playing an important role in stabilizing the growth, promoting employment and improving the people's life. To develop the Internet enterprises healthily is not only the goal of the entrepreneurs, but also the requirement of the country's development. China attaches importance to the dominating role of enterprises in the Internet development. We encourage, support and regulate their development and guide and manage them in accordance with laws. We strengthen the construction of laws concerning the Internet, speed up the establishment and improvement of the law and regulation systems concerning the Internet, strengthen Internet governance in accordance with laws, and urge the enterprises the fulfill their accountability. Besides, China has formulated a series of polices and measures to support the development of the Internet enterprises. It speeds up the reform in examination and approval, funding and patent systems, breaks the barriers in systems and mechanisms, establishes the unified market, reduces the enterprises' burdens and creates favourable environment for their development. Chinese Internet enterprises attach equal importance to economic and social effects, volunteer to be supervised by the government, undertake social and moral responsibilities, reward society and serve the people, and form the joint force for the government-enterprise cooperation and coordination and for the Internet development.

6. Ensuring both security and development

Cyber security and development are the two wings of one bird and two wheels of a cart. China always takes the proper tackling of the relationship between cyber security and development as the priority among priorities of the work concerning cyber security and informatization. The country insists on unified planning, layout, promoting and implementation, and ensures development through security and vice versa. We have formed the correct outlook on cyber security, which is always taken as a priority in informatization, and set up the cyber security accountability. We have carried out a thorough examination on the key information infrastructure, strengthened self-dependent innovation, security inspection and opening and cooperation, and complemented the shortcomings in security with core technology breakthroughs, while consolidating security through key information infrastructure protection to safeguard complete, dynamic, open, relative and common cyber security, and thus to guarantee security for the rapid development of the Internet.

IV. Trend of Future Development and Suggestions on Policies

From October 18 to 24, 2017, the 19th CPC National Congress was held. General Secretary Xi Jinping, on behalf of the CPC Central Committee, gave a report at the Congress. In the report, he analyzed the situation of the world and China today, put forward the Two Centenary Goals and the new "two-step" strategy, described the general blueprint of the socialist modernization, and made overall, strategic and prospective plans, opening up a new era for the socialism with Chinese characteristics. The report also makes strategic planning for promoting the Internet, big data, artificial intelligence and digital economy as well as for constructing a strong Internet country, a digital country and a smart society, providing fundamental principles and action guidelines for the country's Internet development. The coming 30 years will be a great strategic period of opportunity for China's Internet development. At the new historic starting point, the Internet development of China should be guided by Xi Jinping Thought on Socialism with Chinese Characteristics for a New Era, the spirit of the 19th CPC Congress and the people-centered approach. We will speed up the implementation of the strategy on developing and strengthening the country through the Internet, promote the construction of digital China, and ensure more complete and balanced, and fuller and better sustainable development of the country, so that the Internet can better benefit the country and its people, and thus it will contribute more to the modernization of socialism with Chinese characteristics.

1. The Internet development should be shifted from coverage acceleration to quality improvement, and the Internet should be a new facility and platform for economic and social development, with the construction of "digital China" as the leading part

The Internet in China, after 23 years of rapid development, is now experiencing the shift of changes of capacity to changes of quality. At present, the difference between provinces in the field of the Internet is decreasing, the increase of the number of users is not so dramatic, and the social investment in the Internet is becoming rational. There is limited sustainability in the rapid growth resulting from the bonus of facilities, population and capital. In the future, a "Digital China" should be built with high standards, quality and level, the Internet should play a driving and leading role in deepening the supply-side reform, the network infrastructure should be completely upgraded, the promotion of the commercial layout of the new-generation information and telecommunication technology should be accelerated, the next-generation

Internet construction should be sped up, and IPv6 should be laid out in an all-round way. We should make the network infrastructure fast, integrated, intelligent and open, set up infrastructure platforms, data platforms and resource platforms which can improve the economic development quality and efficiency to push the development of the primary, secondary and tertiary industries into the top of the global value chain.

2. Artificial intelligence has become the new "blue sea" of the network information technology. We should occupy the commanding point by taking the R&D of key and basic software and hardware as the focus

In the new active period of the global information technology development, cloud computing, big data, Internet of things, artificial intelligence and other new-generation information technologies are developing fast. In particular, artificial intelligence, as a cutting-edge comprehensive discipline integrating computer science, statistics, cerebral neuroscience and social sciences, will be the leader of the next wave of the world's science and technology revolution as well as of the in-depth technical revolutions in all industries. It will bring about big leap in technology, data and industries. Technically, China's high-performance computing capacity is increasing fast, its deep learning algorithms is evolving, and its mobile Internet is accelerating. The leap of these comprehensive technical capacities makes the breakthroughs in artificial technology possible, so we should concentrate our effort on the R&D of basic computing, chip/processor technology, intelligent operation system, computer vision and image technology, high-speed computing and large-scale data processing technology, and intelligent robots, which are all related to the development of artificial intelligence technology, occupying the commanding point of the new-round information technology development. Industrially, China's mass user data will become the absolute advantage of the artificial intelligence development. Data computing, processing and application of artificial intelligence will create more high-value data for society. We should shift the investment focus from "Internet+" to "artificial intelligence+", launch a new-round innovations and new business and a number of phenomenal artificial intelligence products meeting the requirement of multiple applications and intelligent service to boost innovations and revolutions in industrial organization pattern, service model and business model of different industries.

3. Digital economy is developing fast, so we should make use of the trend of the new-round economic globalization to expand China's digital economy development space

China's digital economy development speed and level are in the leading position of the world, and such economy boasts the potential for sustainable growth. In the future, with the deep integration of the Internet, big data, artificial intelligence and the real economy, a new-round economic globalization driven by data will accelerate, globalization of data trade, e-commerce, Internet finance and online service will be unavoidable, cross-border platform trade volume, demand volume and user number will be increasing by leaps and bounds, more small and medium-sized enterprises and individuals will share the benefits of economic globalization, and platform economies will replace traditional multinationals and become the most influential international trade platforms. China should seize the opportunity to speed up the construction of the new-type open digital economy system and mechanism that combines advantages from both home and abroad and benefits the world. The country will promote intelligent agriculture, manufacturing, online service and Internet finance, and form new growth point and new drive in mid and high-end consumption, sharing economy, modern supply chain, and human capital service. It will lose no time in studying and formulating international rules and standards in terms of cross-border data trade, e-commerce, service trade, and Internet finance. It will also promote the establishment of the Global Digital Economy Development Union, guide Chinese Internet enterprises in participating in the construction of the Belt and Road Initiative, free-trade ports and multi-lateral free trade zones, and further expand the space for its digital economy development and thus improve the development level.

4. Paid content will become the new hot spot of the Internet media profit growth, and we should promote creative transfer and innovative development of Chinese culture

With the increase of cyberspace copyright awareness and the formation of the digital content consumption habit, paid online subscription will become more popular, involving news, video, literature and socializing and other areas. It enables original content and quality content to be sold in a more convenient and efficient way, which, in turn, will lead to a bigger digital content production, service and consumption market. "Content-priority" is becoming more important, so traditional media can make use of content production to find new opportunities for transformation and breakthrough and enhance the influence and yield rate of the content. New media can explore the areas with professional subdivision by optimizing user experience to occupy exclusive copyright resources and constantly increase the content "premium". Chinese culture, with a history of over 5,000 years, combining revolutionary culture and advanced socialist culture, is becoming increasingly popular in the international community and among the people of the world, so we should take advantage of new network technologies and commercial patterns like qualified platforms, artificial intelligence and paid subscription to promote the creative transformation and innovative development and international communication of Chinese culture. We should tell the Chinese story properly to show an authentic, three-dimensional and overall China and thus improve the country's soft power.

5. The construction of a smart society is closely related to the reform effect of the Party and State leadership system and mechanism, and we should promote data resources'opening up and utilization to enhance the modernization of the national governance system and capacity

With the development of the mobile Internet, the popularity of smart phones and the villagers' access to telecommunication service, the construction of a smart society will accelerate, online service and life, and platform governance will become the new patterns of economic and social development, which will produce profound influence on the construction of politics, economy, culture, society and ecology. In the future, e-government will be more intensive, and integrated e-government platforms throughout the country will become the main stream. Therefore, it is necessary to promote the opening up and sharing of data resources, break data barricades between governmental agencies, between network platforms, and between governments and network platforms, and to realize cross-level, cross-region, cross-sector and cross-system governmental administration and coordinated governance. We will keep deepening the development and unitization of governmental information resources and shift the static storage and simple application to dynamic management and deep analysis.

We should find the hidden risks and hazards, fix the bug in supervision and improve the social governance effect by using big data and artificial intelligence. The construction of a smart society will be expanded from individual pilot areas to the whole country. We will make social governance and public service more personalized, intelligent and precise to meet the people's diversified demands and their longing for a better life.

第 1 章　网络基础设施全面提升

网络基础设施是互联网的“大动脉”，具有巨大外部效应，网络基础设施的每一次技术升级、带宽扩容，都会带来互联网普及程度的一次跃升，带来互联网产业应用的新一轮发展。近年来，中国深入实施“宽带中国”战略，政府和行业从多方入手，共同推进中国互联网基础设施加速向高速率、全覆盖、广普及、智能化发展，宽带接入能力显著提升，建成全球最大的 4G 网络；互联网关键资源拥有量大幅增长，利用效率逐步提升，网络服务质量、应用水平和产业支撑能力逐步实现与世界先进水平同步；大力开展“提速降费”行动，农村及偏远地区宽带电信普遍服务补偿机制逐步完善，地区间数字鸿沟明显缩小。

1.1　互联网基础设施成为国家战略性公用基础设施

1.1.1　骨干网络持续优化演进

1. 骨干网顶层架构优化演进，网间互联质量持续提升

自 1994 年全功能接入互联网以来，中国持续加大骨干网建设力度，形成了覆盖全国所有城市、由三家经营性单位（中国电信、中国联通、中国移动）和四家非经营性单位（教育网、科技网、经贸网和长城网）建设运营的

多张高性能骨干网。自 2000 年以来，逐步建立了以北京、上海、广州三个骨干直联点为主、若干交换中心为辅的骨干网网间互联顶层架构。2013 年，新增成都、武汉、西安、沈阳、南京、重庆和郑州 7 个骨干直联点。网间互通带宽持续提升、互联质量不断提升、互通价格持续下降，新增骨干直联点开通后，网间访问路由平均减少 3 跳以上，时延降低 60%～80%，丢包率降低 60%～90%，中西部地区流量绕转现象改善明显，互联性能提升 60%以上。2016 年又新增杭州、贵阳和福州 3 个骨干直连点。骨干网间互通效率明显改善，区域均衡格局基本形成，为各类信息服务应用规模扩大奠定了良好基础。

2. 骨干网容量不断提升，向宽带化、扁平化方向发展

中国骨干网中继带宽经历从 Kbps 到 Mbps、Gbps 的高速提升。自 2000 年以来，伴随光纤工艺、传输、路由设备的多次技术升级，中国骨干网的中继光缆长度增至近 100 万公里，骨干网宽带化进程得到加速，最高单通道通信速率从千比特每秒（Kbps）、2.5G 提升至 10 吉比特每秒（Gbps）、40G 乃至 100G，互联网骨干网带宽实现了每年 50%～100%的高速增长。互联网网络结构持续优化，骨干网和城域网不断扁平化，从星形向网状网演进，网络可控能力逐步增强，骨干网络的疏导效率和用户服务能力大幅提升。

3. 网络互联互通部署初见成效，国际互联能力提升

中国互联网国际出口带宽增长迅速。至 2014 年年底，中国依托位于边境城市的 22 个信道出入口，已与周边 14 个国家和地区实现跨境光缆连接，网络通达亚、美、欧、非、澳等全球各方向，海外网络服务提供点（POP）建设规模达到 72 个。“一带一路”倡议提出后，中国加强与沿线国家信息化互联互通，积极推进中国—缅甸、中国—尼泊尔、中国—巴基斯坦、中国—

吉尔吉斯斯坦等跨境光缆项目建设，筹划建设亚洲、欧洲和美国等方向新海缆和新的国际通信业务出入口。中国联通在沿线热点区域建成 44 个 POP 点[1]，中国移动建成 29 个沿线 POP 点[2]，中国电信建成 67 个 POP 点[3]，与全球主要电信运营企业基本实现对等互联。2010—2015 年，中国国际互联网出入口带宽实现年均速率增长 15%，2016 年增速为 23.1%，达到 6 640.291Gbps[4]。1997—2016 年中国国际互联网出口带宽如图 1-1 所示。

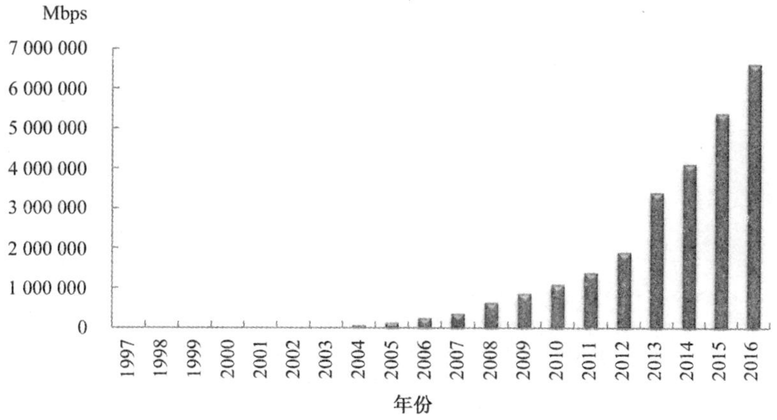

图 1-1　1997—2016 年中国国际互联网出口带宽

（数据来源：CNNIC）

---

[1] 中国联通：四大举措积极助推“一带一路”建设，http://finance.sina.com.cn/roll/2017-05-11/ doc-ifyfecvz 0909616.shtml.

2 中国移动参与“一带一路”共建情况，http://news.163.com/17/0512/13/CK88823E00014 AED .html# from=relevant.

3 中国电信践行“一带一路”倡议·建设篇：全球布局 建设“信息丝路”，http://www.sohu.com/ a/139552103_354877.

4 数据来源：CNNIC，第 39 次《中国互联网络发展状况统计报告》。

4. 下一代互联网建设逐步开展，已进入商用部署起步阶段

中国下一代互联网（NGI）的发展经历了初步引入、应用示范、商用试点、全面部署建设四个阶段，IPv6 升级改造正在积极推进。1998—2001 年，我国先后建成第一个 IPv6 试验床 CERNET-IPv6（1998 年）、第一个下一代互联网交换中心 DRAGONTAP（2000 年）、NSFCNET 试验网（2001 年）等，实现与部分国际下一代互联网互联；2003 年，“中国下一代互联网示范工程（CNGI）”正式启动，核心网包括覆盖全国 22 个城市的 59 个节点及北京、上海两个国际交换中心；2013 年 8 月，启动“国家下一代互联网示范城市”建设工作，全国有 20 多个重点城市全面推进了网络及业务平台的升级改造，大力发展 IPv6 宽带接入用户，但整体进展低于预期，主要原因：一是我国互联网内容提供商（ICP）对 IPv6 支持率低，二是我国 IPv6 商用网络覆盖及用户访问量非常低。2017 年，中国发布《信息通信行业发展规划(2016—2020 年)》及《信息通信行业发展规划物联网分册(2016—2020 年)》，提出到“十三五”期末，国内主要商业网站、教育科研网站和政府网站支持 IPv6，在手机应用排名前 100 的中文版 APP 用户中 80%支持 IPv6，IPv6 流量占比达到 5%。中国下一代互联网正式进入大规模部署和应用阶段，将对未来中国的物联网、车联网、人工智能等新一代信息技术产业发展产生重大促进作用。

1.1.2 宽带网络实现全国覆盖

1. 固定宽带接入技术持续升级，光纤接入已成主流

中国持续加大网络建设投资，自 1995 年开展拨号上网业务以来，从窄带接入低速宽带再到高速光纤，实现了固定宽带接入技术的快速演进升级。2012 年 7 月，国务院印发《“十二五”国家战略性新兴产业发展规划》，提出实施“宽带中国”工程，要求到 2015 年城市和农村家庭分别实现平均 20 兆

和 4 兆以上宽带接入能力。2013 年 8 月，国务院发布《“宽带中国”战略及实施方案》，明确提出宽带网络是新时期中国经济社会发展的战略性公共基础设施，“光进铜退”步伐全面提速，光纤接入（FTTx）成为主流接入技术，中国宽带进入新的发展阶段。截至 2017 年 6 月，互联网宽带接入端口数量达 7.39 亿个，光纤接入 FTTH/0 端口达到 5.99 亿个，占互联网宽带接入端口总数比重由上年末的 75.6%提高到 81%。中国固定宽带用户数超过 3.22 亿户，其中光纤接入用户达到 2.61 亿，占固定互联网宽带接入用户总数的 80.9%[5]。2006—2017 年 6 月中国固定宽带用户发展情况与国际比较如图 1-2 所示。

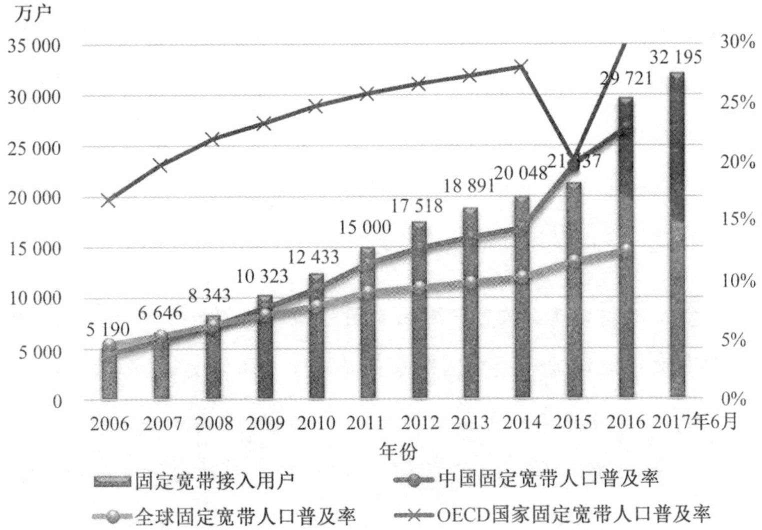

图 1-2　2006—2017 年 6 月中国固定宽带用户发展情况与国际比较

（数据来源：工业和信息化部，世界银行，OECD）

5 数据来源:工业和信息化部，《2017 年 1—6 月份通信业经济运行情况》，http://www.miit.gov.cn/n1146285/n1146352/n3054355/n3057511/n3057518/c5734845/content.html.

专栏 1　中国三大电信运营商

历经数次电信领域改革,目前中国电信运营市场由中国移动、中国联通、中国电信三家公司主导。三大电信运营商是中国网络基础设施建设的主力军。

中国移动（全称“中国移动通信集团公司”）于 2000 年 4 月 20 日成立，是一家基于 GSM，TD-SCDMA 和 TD-LTE 制式网络的移动通信运营商。中国移动通信集团公司是在原中国电信移动通信资产总体剥离的基础上组建的国有骨干企业，经营规模居三大运营商之首。中国移动通信集团公司全资拥有中国移动（香港）集团有限公司，由其控股的中国移动有限公司在国内 31 个省（自治区、直辖市）和香港特别行政区设立全资子公司，并在香港和纽约上市。截至 2017 年 10 月，市值达到 1.6 万亿元人民币。

中国联通（全称“中国联合网络通信集团有限公司”）于 2009 年 1 月在原中国网通和原中国联通的基础上合并组建而成，在国内 31 个省（自治区、直辖市）和境外多个国家和地区设有分支机构。中国联通主要经营 GSM、WCDMA 和 FDD-LTE 制式移动网络业务及固定通信业务，网络接入业务和各类电信增值业务等。中国联通是中国唯一在纽约、香港特别行政区、上海三地同时上市的电信运营企业。截至 2017 年 11 月，市值达到 2 390 亿元人民币。

中国电信（全称“中国电信集团公司”）成立于 2000 年 9 月，在全国 31 个省（自治区、直辖市）和美洲、欧洲、中国香港、中国澳门等地设有分支机构，拥有覆盖全国城乡、通达世界各地的通信信息服务网络，建成了全球规模最大、国内商用最早、覆盖最广的 CDMA 3G

网络。中国电信股份有限公司于 2002 年在香港和纽约上市，中国通信服务股份有限公司于 2006 年在香港上市，截至 2017 年 11 月，市值达到 2 680 亿元人民币。

2. 移动通信网络实现 4G 基本覆盖

1994 年，中国开通第一部 GSM 移动电话，标志着中国移动通信技术由模拟时代进入 2G 数字时代。2009 年，中国发放 3G 牌照，并在当年迅速建成三个在全球规模领先的 3G 网络，开创了全球电信发展史上建设规模最大、速度最快的新纪录。2010—2013 年，中国 3G 用户大幅增长，移动互联网接入流量年均复合增长率超过 50%，3G 逐步替代 2G 成为移动宽带流量的主要承载网络。2012 年 1 月 18 日，中国主导的时分长期演进技术（TD-LTE）被国际电信联盟确定为第四代移动通信国际标准之一。2013 年 12 月和 2015 年 2 月，第四代移动通信（4G）牌照分两批发放，中国建成了世界最大的 4G 网络，2G/3G 用户加速向 4G 迁移。截至 2017 年 6 月，中国移动通信用户总规模超过 13.6 亿户，其中 4G 用户达到 8.9 亿户，占移动用户总数的 65.1%[6]。2009—2016 年中国移动互联网接入流量规模及用户月均流量如图 1-3 所示。2009—2016 年中国新增移动通信用户构成如图 1-4 所示，2009—2017 年 6 月中国移动通信用户及新增移动通信用户规模如图 1-5 所示。

6 数据来源:工业和信息化部，《2017 年 1—6 月份通信业经济运行情况》，http://www.miit.gov.cn/n1146285/n1146352/n3054355/n3057511/n3057518/c5734845/content.html.

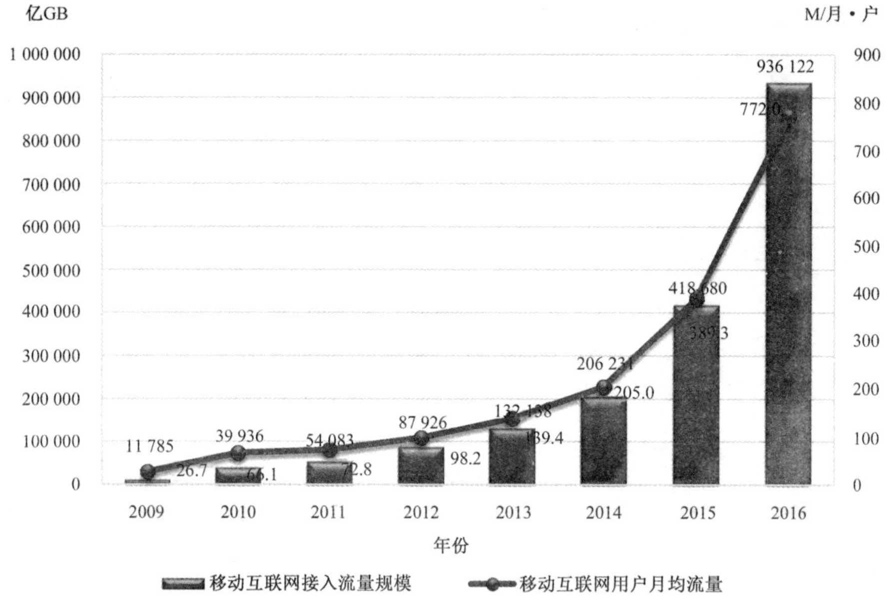

图 1-3 2009—2016 年中国移动互联网接入流量规模及用户月均流量

（数据来源：工业和信息化部）

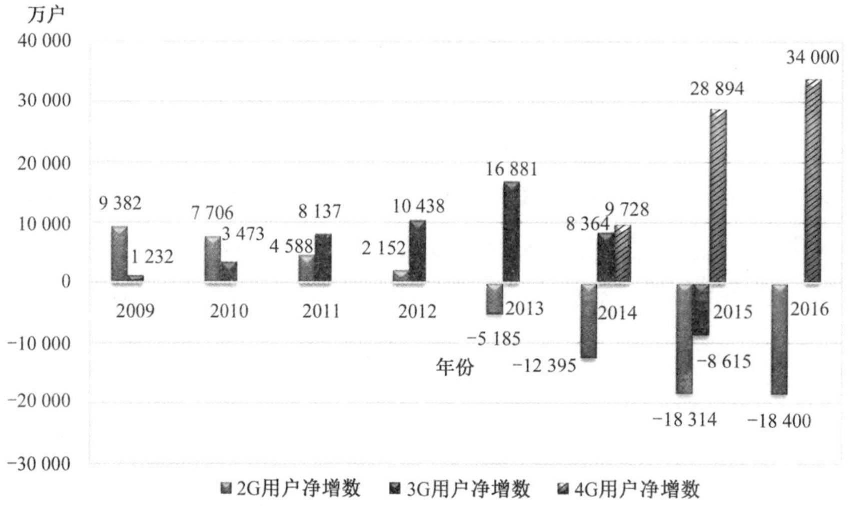

图 1-4 2009—2016 年中国新增移动通信用户构成

（数据来源：工业和信息化部）

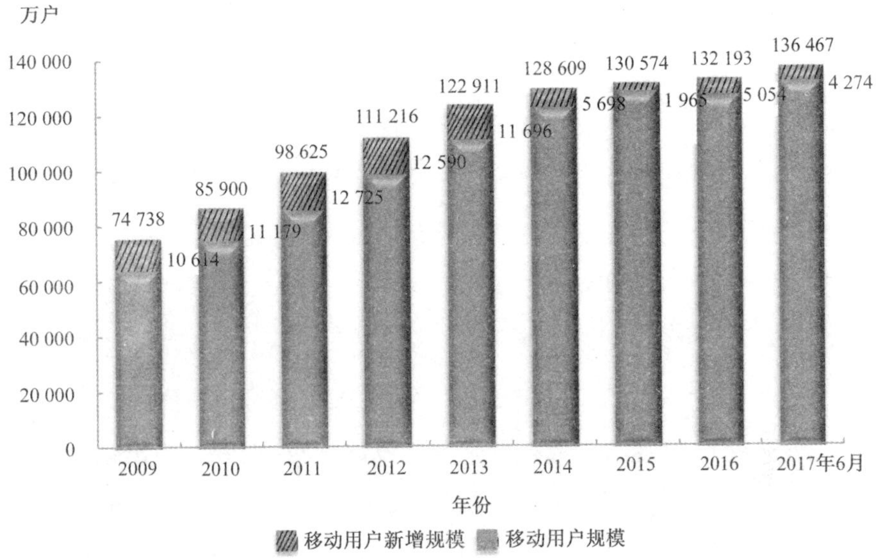

图 1-5　2009—2017 年 6 月中国移动通信用户及新增移动通信用户规模

（数据来源：工业和信息化部）

3．5G 移动网络加速到来

中国高度重视 5G 技术发展，早在 2013 年就成立由“产、学、研、用”各方组成的 5G 推进组（IMT-2020 推进组），充分发挥平台优势，推动 5G 技术研发、验证技术方案、支撑国际标准的制定。自 2016 年以来，中国政府先后发布《中华人民共和国国民经济和社会发展第十三个五年规划纲要》《“十三五”国家信息化规划》和《国家信息化发展战略纲要》，均对积极推进 5G 发展作出部署。按照《信息通信行业发展规划（2016—2020 年）》部署，中国将于 2018 年进行大规模 5G 试验组网，并在此基础上于 2019 年启动 5G 网络建设，最快在 2020 年正式推出商用服务。

4."提速降费"不断深化

2015 年 5 月，国务院常务会部署加快建设高速宽带网络促进提速降费的措施，助力创业创新和民生改善，包括确定加快高速宽带网络建设，推进光纤到户和宽带乡村工程，推进电信市场开放和公平竞争，移动转售业务和宽带接入网业务试点工作全面开展。2015—2016 年，中国移动向客户让利超过 560 亿元，中国电信手机流量费下降 57.4%，中国联通固定宽带资费一年降幅超过 67%[7]。2017 年是我国深入实施"提速降费"的第三年，三大运营商推出流量不清零政策，不断降低流量资费，下调中小企业专线资费以及国际热门、重点方向长途资费等。截至 2017 年 6 月，全国 20M 以上宽带用户比例达到 86%、4G 用户占比达到 65.1%[8]，提速降费取得一系列实际成效。

5. 电信普遍服务能力大幅提升

中国政府一直高度重视农村信息基础设施发展问题。2004 年 1 月，信息产业部出台《农村通信普遍服务——村通工程实施方案》（以下简称"村通工程"），要求基础电信运营企业采取分片包干、自筹资金、自行经营维护的方式，承担农村及偏远地区的通信网络建设，采用"三个三步走"推进策略——在网络地域覆盖上，从乡镇到行政村再到自然村分三步延伸；在业务提供水平上，从通电话到通互联网再到通宽带逐级提升；在信息化发展进程上，从建基础设施到构建业务平台再到推广适农业务层层推进。2014 年，

---

7 新一轮提速降费开启 三大运营商公布的新方案都在这，http://finance.sina.com.cn/chanjing/cyxw/2017-03-06/doc-ifycaasy7719405.shtml.

8 2017 年 6 月通信业主要指标完成情况，http://www.miit.gov.cn/n1146312/n1146904/ n1648372/c5735472/content.html.

国家发展改革委、财政部、工业和信息化部联合启动实施“宽带乡村”工程，利用多渠道资金大力支持农村宽带建设和应用普及，当年年底共完成 3 000 余个乡镇和 15 万多个行政村宽带建设，1.8 万个特困村实现互联网覆盖、6 400 个农村学校和公益机构通宽带。2015 年 10 月，国务院常务会议明确实施“改革创新电信普遍服务补偿机制”，对农村及偏远地区宽带投资采取由中央、地方和企业共同推进的多元化资金支持和市场化运作机制，这是中国普遍服务机制的一次重大政策突破，为实现 2020 年农村宽带发展战略目标提供了有力支撑。截至 2017 年第三季度，工业和信息化部联合财政部组织实施电信普遍服务两批试点，各方投资超过 400 亿元，共部署 13 万个行政村通光纤，其中有 4.3 万个是建档立卡的贫困村，电信普遍服务试点实施使我国农村和偏远地区网络的发展不断加快，条件得到了极大的改善，中国行政村通宽带的比例已经超过了 96%，贫困村宽带的覆盖率已经达到了 86%[9]。

1.1.3　应用基础设施有力推动互联网业务发展

1. 内容分发网络覆盖日益广泛，分发能力持续提升

从 1998 年开始，内容分发网络（CDN）产业在中国大致经历了初创试水、稳步发展和加速发展三个阶段。近年来，内容分发网络呈现加速发展态势，涌现多家专业担负内容分发网络服务的互联网企业，各主要基础电信运营商已将内容分发网络作为增值业务进行重点布局，网络覆盖范围不断扩

9 http://www.miit.gov.cn/n1278117/n4310819/n4310832/c5882939/content.html.

大、内容分发能力不断提升、承载业务范围不断拓展。到 2014 年，中国已建成各类内容分发网络节点数超过 2 400 个，峰值带宽储备超 10TB[10]。2016 年 12 月，工信部向网宿科技和阿里云颁发内容分发网络业务经营许可证，标志着内容分发网络业务进入专业许可运营阶段。截至 2017 年 6 月，中国共有 16 家企业获得内容分发网络牌照[11]，在国内外部署数千个节点、均具备 T 级带宽规模。阿里巴巴、百度、腾讯、网宿、蓝汛、盛大等企业纷纷基于内容分发网络，提供云存储、云安全、云监控、云数据库、云主机、云计算等服务，内容分发网络进一步成为云服务分发落地的重要平台。

专栏 2　内容分发网络（CDN）名词解释

CDN 的全称是 Content Delivery Network，即内容分发网络。其基本思路是尽可能地避开互联网上有可能影响数据传输速度和稳定性的瓶颈和环节，使内容传输更快、更稳定。通过在网络各处放置节点服务器所构成的在现有的互联网基础之上的一层智能虚拟网络，CDN 系统能够实时地根据网络流量和各节点的连接、负载状况以及到用户的距离和响应时间等综合信息将用户的请求重新导向离用户最近的服务节点上。其目的是使用户可就近取得所需内容，解决 Internet 网络拥挤的状况，提高用户访问网站的响应速度。

2. 互联网数据中心规模快速增长，云计算服务成为重要发展方向

随着中国互联网产业的不断壮大，数据中心从小规模简单存储计算中心转向以支持网络信息服务为核心的互联网数据中心（IDC），功能从简单的“数据仓库”发展成为集数据交互、信息处理和数据分析的综合服务。2004

10 资料来源：《内容分发网络（CDN）白皮书（2014 年）》，原工业和信息化部电信研究院。

11 工信部发放第六批 CDN 牌照 获牌企业增至 16 家，http://www.sohu.com/a/164171420_488920.

年以来，各基础电信运营企业加大对数据中心建设的投入，数据中心建设快速向规模化、集中化、绿色化演进，市场规模保持两位数增长。2013 年 1 月，《关于数据中心建设布局的指导意见》发布，主要互联网企业、基础电信运营企业以及其他行业的部分大型企业积极推进云计算数据中心建设，云计算资源与服务的承载和调度能力大幅提升。2012—2014 年，中国数据中心市场规模从 2011 年的 80.4 亿元增长到了 2014 年的 156 亿元，数据中心总量超过 40 万个。其中，经营性数据中心机房约 1 000 个、机柜约 28 万架、服务器超过 300 万台[12]。截至 2016 年 10 月底，工信部已颁发 327 张跨省数据中心业务经营许可证、844 张省内经营许可证[13]。

1.2　互联网基础资源保有量大幅增长

1.2.1　IP 拥有量上升至世界第二位

1. IP 地址拥有量快速增长，有效利用率仍然偏低

从 20 世纪 90 年代初中国从亚太网络信息中心（APNIC）申请到第一个 IPv4 地址开始，在互联网业务发展的推动和各方共同努力下，中国 IP 地址资源规模发展迅速。截至 2017 年 6 月，中国拥有 IPv4 地址数量 3.38 亿个，IPv6 地址数量 21 283 块/32 地址[14]，均居世界第二位，但仍不能有效满足发

12 数据来源：赛迪智库。

13 2016 年 IDC 服务大会在京成功召开，https://www.csdn.net/article/a/2016-12-28/15842846.

14 数据来源： CNNIC，第 40 次《中国互联网络发展状况统计报告》。

展需要。一方面，IP 资源人均占有水平较低，随着中国网民规模增加、网络应用水平提高、网络消费能力扩大，对网络服务质量的要求也日益提高，IP 资源与互联网发展的不匹配现象会更加明显；另一方面，IPv4 地址已分配完毕，而 IPv6 仍未在我国得到大规模有效利用。根据 APNIC 研究数据，截至 2016 年年底，中国 IPv6 用户近 500 万（见图 1-6），普及率仅为 0.7%、居全球第 56 位，而同期普及率最高的国家是比利时，已超过 60%，其次是美国，普及率为 34%。

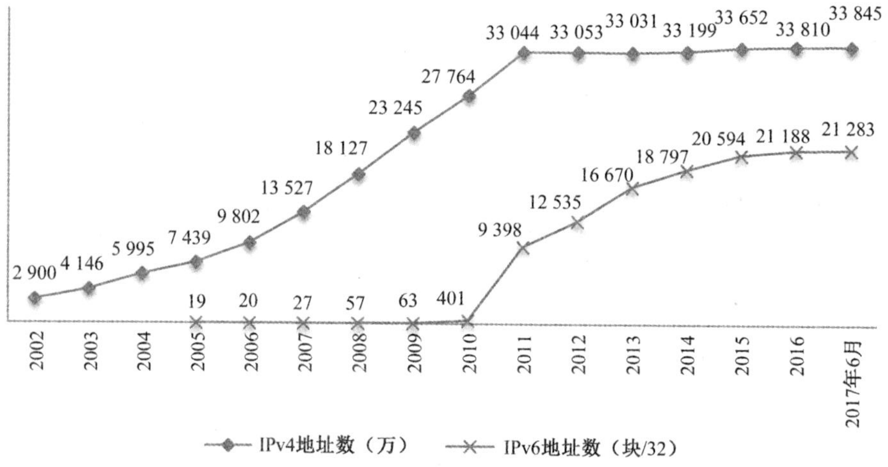

图 1-6　中国 IPv4 与 IPv6 数量（2002—2017 年 6 月）

（数据来源：CNNIC，中国互联网络发展状况统计报告）

1.2.2　域名注册数量达千万量级

1. 中国域名注册数量突破 2 000 万，成为世界第一大国家顶级域名

1994 年 5 月，在钱天白教授和德国卡尔斯鲁厄大学的协助下，中国科

学院计算机网络信息中心完成了中国国家顶级域名（.cn）服务器的设置。经过多年发展，中国域名注册数量实现了快速增长，域名解析基础设施不断升级。2003年国家顶级域名（.cn）向公众开放注册，“.cn”域名获得了迅猛发展；2008年7月，“.cn”域名注册量达1 218.8万个，首次成为全球第一大国家顶级域名；截至2016年年底，中国域名总数达到4 228万个，年增长达36.3%（见图1-7）。其中“.cn”域名注册数量达到了2 061万个，年增长达25.9%[15]。

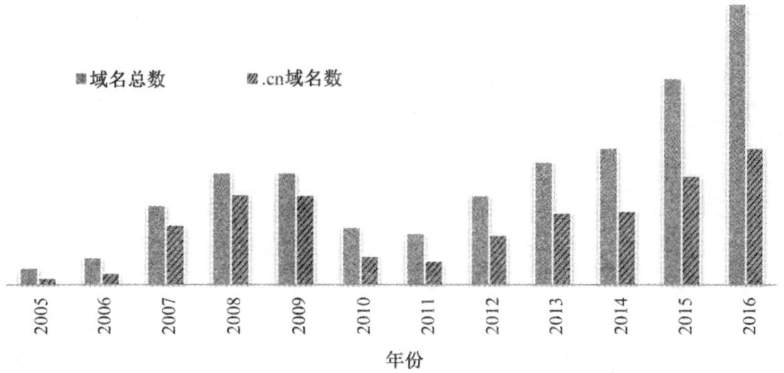

图1-7　中国域名总数和.cn域名数（2002—2016年）

（数据来源：CNNIC，中国互联网络发展状况统计报告）

2. 中文域名大力发展，多个中文顶级域名纳入全球互联网域名体系

1998年，中国互联网络信息中心（CNNIC）开始中文域名技术研发工作，揭开了汉字字符融入国际互联网根域名系统的序幕。2010年，中文国家顶级域名“.中国”和“.中國”入根，并实现全球解析。2012年新通用顶

15 数据来源：CNNIC。

级域名开放申请以来，“.网络”、“.公司”、“.公益”、“.政务”等多个中文顶级域名成功实现入根和全球解析。自 2014 年 7 月开始，中文新通用顶级域名“. 公司”、“. 网络”正式开放注册。截至 2016 年 12 月，“.cn”域名数达到 2 061 万个，占总域名数的 48.7%；“.中国”域名数达到 47.4 万个，占总域名数的 1.1%[16]。2016 年中国分类域名数见表 1-1。

表 1-1　2016 年中国分类域名数

项目	数量（个）	占域名总数比例
CN	20 601 491	48.7%
COM	14 345 243	33.9%
NET	1 633 071	3.9%
ORG	330 457	0.8%
中国	474 115	1.1%
BIZ	210 062	0.5%
INFO	224 321	0.5%
其他	4 456 942	10.5%
合计	42 275 702	100.0%

3. 丰富域名资源和优质解析服务，催生海量网站设立

自 1997 年起，中国网站数量从零开始快速增长，仅 2000 年，网站数量从 1.5 万增长至 26.5 万。伴随着互联网产业的壮大，2009 年网站数量突破 300 万个。2010 年，中国政府加强互联网领域安全治理，网站数减少到 191 万个，但网页数和网页字节等仍然保持大幅度增长，网站质量随着“水分”的挤出而得到提升。截至 2017 年 6 月，中国网站总数达到 506 万个[17]，如图 1-8 所示。

16 数据来源：CNNIC。

17 数据来源：CNNIC。

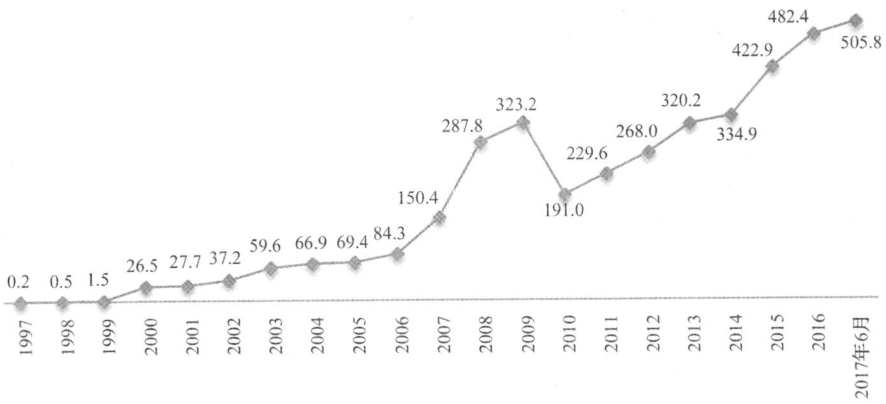

图 1-8　中国网站数（万个）（1997—2017 年 6 月）

（数据来源：CNNIC，中国互联网络发展状况统计报告）

1.2.3　解析服务不断升级支撑网站服务

1. 积极提升域名解析服务质量，建成高效域名解析服务体系

实现中国互联网全面联通世界，不仅需要“.cn”域名解析，对世界其他顶级域名的解析也同样十分重要。中国从 2003 年拥有第一个根服务器镜像开始，始终积极引入根服务器的镜像节点和“.com/.net”顶级域镜像节点，截至 2017 年，已开设 F、I、J、L 四个互联网域名根服务器镜像节点（见表 1-2）。同时，中国在全球设置域名解析服务节点，形成了境内覆盖所有网络运营商，境外覆盖亚太、欧洲、北美等地区的解析服务设施，为全球用户访问中国网站提供良好服务，不断改善用户体验。自此，我国较为全面的域名解析基础设施已基本完成建设。

专栏 3　镜像服务器名词解释

镜像服务器（Mirror Server）与主服务器的服务内容都是一样的，只是放在一个不同的地方，分担主机的负载。简单来说，类似照镜子，镜子里的影像和本人完全一致。在网上内容完全相同而且同步更新的两个或多个服务器，除主服务器外，其余的都被称为镜像服务器。

表 1-2　国内引进根镜像部署及政府监管情况

根	部署地点	引入时间	引入单位	根管理机构
F	北京	2003 年	中国电信	ISC
F	北京	2003 年	科技网/CNNIC	ISC
I	北京	2005 年	科技网/CNNIC	Netnod
J	北京	2006 年	联通（原中国网通）	Verisign
L	北京	2012 年	科技网/CNNIC	ICANN

第 2 章　信息技术取得重大突破

中国始终把技术创新摆在突出位置，深入实施创新驱动发展战略，注重发挥互联网企业等各类创新主体作用，集中力量开展互联网关键技术、共性技术研究，在集成电路、基础软件、核心元器件、通信技术、云计算、大数据、物联网、下一代互联网、卫星导航等多个领域取得重大突破，为实现由网络大国向网络强国的跨越提供有力支撑。

2.1　通信技术发展与世界同步

经过 20 多年的发展，中国在通信技术领域实现从追赶到比肩再到领先的巨大跨越，在网络设备技术、国际通信标准制定等领域跨入全球领先行列，在量子通信研究、空间导航技术等领域取得重大突破，在数据通信技术国际专利申请量方面位居全球前列，在网络设备产业、移动终端制造等方面实现了由点到面的集体突破，涌现出了华为、中兴等一批世界级通信技术企业。

1. 通信技术从整体跟随到局部突破，逐步实现重点引领

自 1994 年实施信息化重点工程和电信网络基础设施建设工程以来，我国网络技术研究与通信设备研制开始整体跟随国际先进技术，并在光通信、数据通信等重点领域率先取得突破，依赖进口情况明显改观。1995 年，我国研制出第一套 155Mb/sSDH 设备和 ATM 交换机，打破了国际设备垄断国内市场的局面，光通信领域发展迅猛，已成为通信技术、设备、管理体系的

重要输出国。2006 年，由烽火科技为主承担的“40Gb/s SDH 光纤通信设备与系统”通过国家验收，成为国际上第一套符合 ITU-T 相应标准的光纤通信设备。华为于 2013 年发布电信级 SDN 控制器 SNC，2016 年发布的 32 波 10G WDM-PON（Wavelength Division Multiplexing Passive Optical Network）样机支持单根光纤 320G 总带宽，2017 年与中国电信联合完成全球首个大容量路由器关键技术（400GE）中长距技术研发与综合评测，这些设备与技术均属全球首发。

2. 积极参与国际通信技术标准制定

随着网络技术创新能力和产业化能力的快速提升，中国在国际标准制定中日益活跃。在国际电联（ITU）方面，中国自 2000 年提交第一个 IP 国际标准（ITU-T X.85）并获得批准后，在相关领域的话语权不断提升，截至 2017 年 11 月，中国向 ITU-T 提供的标准文稿数达 500 多个[1]。在互联网工程任务组（IETF）方面，从 1996 年的 RFC1922 开始到 2008 年，中国主导完成的 RFC 共有 7 项，从 2013 年到 2017 年 10 月，在 IETF 主导或参与完成的标准达到 185 项[2]，后发动力强劲。我国多个部门、企业、院校成为 ITU-T、IETF 等国际标准化组织的重要成员，并有多位中国专家担任领导职位。这有助于推动中国技术和标准“走出去”，推进我国网络设备供应企业和产业的国际化发展。

3. 数据通信技术创新能力增强

伴随着网络基础设施的大规模建设和互联网应用的快速普及，中国数据

1 数据来源：ITU 中国官方网站，http://www.ituchina.cn/itut/zgwg/.

2 IETF 官方网站资料收集，https://www.rfc-editor.org/rfc-index-100a.html.

通信技术创新能力显著增强。2012—2016 年，我国通信设备制造企业五年中四度位居全球国际专利申请量第一。中国企业自主研发高端网络处理器芯片（NP）取得成功，并在全球率先发布 T 级路由器，实现了从模仿到超越，从中低端市场迈向高端市场“质”的飞跃。我国网络设备产业由弱到强，实现了由点到面的集体突破，涌现出华为、中兴等一批成长迅速的高科技企业，在固定宽带接入系统和光传输系统上的技术实力居各国前列。在近两年国际通信设备产业整体低迷的情况下，华为、中兴等国内企业凭借集成创新优势，持续保持扩张势头。2017 年上半年，华为销售收入为 2 831 亿元人民币，利润为 311 亿元人民币[3]，继续领跑全球；中兴、烽火等企业也保持两位数高速增长，企业规模均已大幅超越爱立信、阿朗、诺基亚等国际同类企业。

4．移动通信技术实现领跑，5G 技术研发试验全面启动

中国移动通信经历了 1G 空白、2G 跟随、3G 突破、4G 同步的发展历程，正在争取实现 5G 引领的新目标。20 世纪 90 年代，中国引进全球移动通信系统（GSM）、码分多址（CDMA）等国外 2G 技术标准，逐步实现设备和产品的自主研发；2000 年 5 月，中国拥有自主知识产权的时分同步码分多址（TD-SCDMA）标准技术被国际电信联盟（ITU）正式批准成为 3G 国际标准之一，于 2009 年实现成功商用。在 4G 发展上，中国采取技术创新和标准化采取融合战略，将 TD 核心技术与 LTE 国际主流标准融合形成 TD-LTE 标准，2010 年 9 月成为 4G 国际标准，并形成了由 10 家系统、14 家芯片、80 多家终端组成的产业生态。2016 年 1 月，我国 5G 技术研发试验全面启动，2016 年 9 月，已完成第一阶段技术试验，5G 单点关键技术的可行性得到验证；2017 年 9 月，第二阶段无线部分测试结果已发布，各厂商的 5G 技术集成方案均满足国际电信联盟（ITU）制定的峰值速率、时延、用户连接

3 数据来源：企业年报。

能力、流量密度等关键指标，为中国 5G 按期实现商用打下了基础。

5. 光通信领域局部市场实现领先

1995年，中国研制出第一套155兆比特每秒（Mbps）同步数字体系（SDH）设备和异步传输模式（ATM）交换机，打破了国际设备垄断国内市场的局面。近年来，中国企业在 100G 光传输网和密集波分复用（OTN/DWDM）、分组传输网/IP 无线接入网（PTN/IP RAN）、10G 以太网 PON/千兆 PON（EPON/GPON）接入等光传输技术领域取得实质性突破，在 400GB 乃至 1TB 研究领域中国厂商逐渐从追赶到同步，部分领域实现领先。超高速超大容量超长距离光传输、软件定义弹性光网络、下一代光接入网、硅光子集成等前沿技术形成一定技术积累。光纤预制棒等关键器件和材料自给能力显著增强。光通信市场规模不断扩大，产品种类不断丰富，技术实力和产业化能力大幅提升。2015 年，烽火通信等光纤厂商分别宣布在低损耗大有效面积光纤领域取得突破，并发布了适用于 100G/400G 骨干网的产品，在高端骨干网光缆市场占据高位，产品技术已经接近或达到世界领先水平。但中国在光通信芯片领域技术实力仍相对偏弱，中国光器件厂商技术积累与研发实力仍相对薄弱，对光子芯片领域的研发支持和投入仍需不断加强。

6. 量子通信领域实现领先突破

自 2016 年以来，在国家战略规划引导和业界共同努力下，中国在量子通信领域取得一系列重要成果。2016 年 8 月，首颗量子科学实验卫星“墨子号”成功上天，并完成量子纠缠分发、量子密钥分发、量子隐形传态实验。2017 年 5 月，由中国科技大学、中国科学院-阿里巴巴量子计算实验室、浙江大学、中科院物理所等协同完成参与研发的光量子计算机诞生。2017 年 9 月，首条量子保密通信网络“京沪干线”全线贯通。中国量子通信研究在国际上处于领先地位，并在国防、金融、政务专网等重要安全领域分别启动广域量子保密通信骨干网络建设、人民币跨境收付信息管理系统的量子通信技术应用示范等实验项目。

7. 北斗卫星导航系统步入全球组网时代

北斗卫星导航系统是中国自行研制的全球卫星导航系统，是继美国全球定位系统、俄罗斯洛格纳斯卫星导航系统之后全球第三个成熟的卫星导航系统。2000年发射第一颗卫星，2012年实现向亚太大部分地区提供服务，2013年出台的《国家卫星导航产业中长期发展规划》提出“重点推动北斗兼容卫星导航功能成为车载导航、智能手机的标准配置”。根据《“十三五”国家信息化规划》部署，北斗系统建设应用行动从2017年开始实施。2016年8月，在新申请入网的3 500余款手机终端中，支持北斗定位功能的款型占比超过40%[4]。2016年，我国卫星导航与位置服务产业总体产值达到2 118亿元，其中与卫星导航技术直接相关的芯片、器件、算法、软件、导航数据、终端设备等在内的产业核心产值达到808亿元，北斗对产业核心产值的贡献率已达到70%，北斗导航应用已渗透到交通运输、车船监管、电力、农业、渔业、公安、林业等诸多行业和领域[5]。2017年11月，中国成功发射两颗北斗三号卫星，这标志着中国北斗卫星导航系统步入全球组网时代。

8. 海底光缆加速连接世界

海底光缆作为当代国际通信的重要手段，承担了90%的国际通信业务，是全球信息通信的主要载体，已成为我国光纤企业的新业务领域。1989—

4 《中国卫星导航与位置服务产业发展白皮书》，中国卫星导航定位协会。

5 《中国卫星导航与位置服务产业发展白皮书》，中国卫星导航定位协会。

1998 年，中国先后参与了 18 条国际海底光缆的建设与投资。其中，1996 年 2 月，中韩海底光缆建成开通；1997 年 11 月，我国参与建设的第一条在中国登陆的洲际光缆系统——全球海底光缆系统（FLAG）建成并投入运营，中国段为 622 千米；2000 年 9 月，亚欧海底光缆上海登陆站的开通，标志着由中国电信集团公司参与建设，连接亚欧海底 33 个国家和地区的亚欧海底光缆系统正式开通；2017 年 6 月，中国联通、喀麦隆电信、华为海洋共同签署南大西洋国际海底光缆（SAIL）建设协议，建成后将成为非洲和南美洲之间传输容量最大、时延最小的海缆路由。

2.2 集成电路技术实现多环节协同突破

集成电路作为战略性、基础性和先导性产业，是培育战略性新兴产业、发展信息经济的重要支撑，在信息技术领域的核心地位十分突出。为促进集成电路产业发展，2014 年中国出台《国家集成电路产业发展推进纲要》。2017 年 3 月，由 62 家龙头企业、高校、研究院所和社会组织等共同发起的“集成电路产业技术创新战略联盟”在北京成立。在市场拉动和政策的双轮驱动下，我国集成电路技术能力显著提升，集成电路设计技术在部分领域达到先进水平，22nm 技术研发实现突破、28nm 制造工艺实现规模量产，封装技术实现与全球同步。

1. 中国集成电路设计能力进入纳米级水平，设计工具依赖性较强

在计算机辅助设计（Computer Aided Design，CAD）应用发展的带动下，中国集成电路设计能力出现显著进步，已经从逆向设计过渡到正向设计，开

始进入超大规模集成电路设计和片上系统（System On a Chip，SoC）设计阶段，并在部分设计要求较高的电子产品中采用全定制设计。近年来，深圳海思、紫光展锐、中芯国际、展讯通信、中国华大等专业集成电路设计公司崭露头角，进入全球主流 16nm / 14nm 的设计水平。中国在电子设计自动化（Electronic Design Automatic，EDA）工具自主研发取得一定成绩，华大九天公司研发出全套 EDA 工具软件包，但其竞争力仍然落后于世界领先水平，先进 EDA 工具全部掌握在国外公司中。此外，中国集成电路知识产权核（Intellectual Property Core，IP 核）设计技术发展起步晚，目前存在设计种类少、经过验证的 IP 核数量少等问题，且来自国外的 IP 核正在以合作方式向国内企业提供低价甚至免费的 IP 核产品，制约着我国 IP 核自主设计能力和产业化应用的发展，进而影响 SoC 设计能力的构建。

2．集成电路芯片制造技术突破纳米关键工艺

长期以来，中国在集成电路制造技术方面缺乏自主知识产权，导致中国芯片产业一直受到西方在先进制造装备、材料和工艺引进等方面的种种限制，制约着中国在计算机、消费类电子、网络通信、汽车电子等领域的赶超，目前这一局面正在得到改观。2013 年，中科院微电子研究所集成电路先导工艺研发中心在 22nm 关键工艺技术先导研究与平台建设上实现重要突破，推动我国集成电路制造产业自主话语权的建立，为我国自主研发 16nm 及以下技术代的关键工艺提供了技术支撑。目前，中国正在推进 16nm/14nm 制造技术投产，加紧 10nm 技术研发和 7nm、5nm 技术预研。中芯国际是目前中国大陆唯一能提供 28nm 制程服务的纯晶圆代工厂，2016 年中芯国际 28nm 营收占比约 5%，预计 2017 年将达到约 10%[6]。

6 数据来源：企业年报。

3. 封装测试业技术水平正在向国际主流领域进发

在国家科技重大专项“极大规模集成电路制造装备及成套工艺”支持下，中国封装测试业技术能力和工艺水平不断提高，多个高端封装工艺及集成封装技术开发项目顺利通过专项组织验收，40nm / 28nm 等技术节点相应的高密度封装技术、大功率器件封装技术、封装系统集成技术等多种先进封装技术开发成功或规模导入量产，有力促进了集成电路封测产业链的健康发展。我国新型自主知识产权封装技术获得国际认可，截至 2017 年 9 月，中国企业已获得国内/国际专利 585 项，为上百家企业提供数百项封装测试技术服务。“十三五”期间，我国将大力推动倒装芯片、凸点封装、晶圆级芯片封装等封装技术的规模化生产能力，开展硅通孔、3D 堆叠封装等新型封装技术开发。

2.3 大规模计算技术快速追赶全球高水平

中国在云计算、大数据、人工智能等新兴技术领域紧盯世界发展步伐，研发应用活跃。云计算骨干企业在大规模并发处理、海量数据存储、数据中心节能等关键领域取得突破，部分指标已达到国际先进水平。在超大规模数据仓库、分布式存储和计算、基于人工智能的大数据分析等技术方面已达到国际先进水平。人工智能技术推动机器人在感知、控制和人机交互能力的大幅提升，得到快速升级和部署应用，正在成为“互联网+”与“中国制造”的重要力量。

1. 超级计算机实现自主可控，达到国际先进水平

中国的超级计算机研制起步于 20 世纪 60 年代，大体经历了三个阶段：第一阶段，自 60 年代末到 70 年代末，主要从事大型机的并行处理技术研究；第二阶段，自 70 年代末至 80 年代末，主要从事向量机及并行处理系统的研制；第三阶段，自 80 年代末至今，主要从事大规模并行处理系统（MPP）及工作站集群系统的研制。经过几十年不懈努力,我国高端计算机系统研制取得丰硕成果，“银河”、“曙光”、“神威”、“深腾”等一批国产高端计算机系统相继问世，使中国成为继美国、日本之后第三个具备研制高端计算机系统能力的国家。2017 年 6 月，在世界超算大会上，中国自主研发的“神威 •太湖之光”超级计算机系统以每秒 12.5 亿亿次的峰值计算能力及每秒 9.3 亿亿次的持续计算能力，再次名列国际 TOP 500 榜单之首。中国“天河二号”也第三次携手获得前两名。

2. 云计算技术自主研发能力增强，部分关键技术领域取得突破

中国云计算骨干企业在大规模并发处理、海量数据存储、数据中心节能等关键技术领域取得突破，部分指标已达到国际先进水平。2017 年年初，中科曙光公司研发的全球首款亿级并发云服务器系统在天津实现量产，使采用传统架构服务器的数据中心在现有硬件水平下，拥有更大计算能力，以应对海量的并发访问。中国云服务厂商成功研发免重启热补丁技术，如 UCloud 的热补丁技术可实现免重启修复所有内核代码，并将热修复过程业务中断时间控制在 10ms 内，已累计进行了约 5 万台次的热补丁修复。

3. 互联网企业实现技术引领，各类云服务蓬勃发展

基础设备方面，中国企业实现 EB 级存储系统相关软硬件、亿级并发服务器系统等核心技术和产品研发突破。2011 年，百度、阿里巴巴、腾讯共

同启动了“天蝎项目”，积极与基础电信运营企业开展合作，建设形成一体化高密度的整机柜服务器解决方案，占据了国内主要市场，并对国际市场产生积极影响。阿里云、腾讯云等拥有自主知识产权技术体系的云服务平台已经进入国际市场，国内整体产业正在步入快速爆发阶段。截至 2017 年 10 月，阿里云生态已拥有超过 5 000 家合作伙伴，云市场商品逾 6 000 款[7]。

4. 互联网企业成为大数据创新孵化平台

各大互联网企业是中国大数据产业的创新孵化主体。腾讯分布式大规模数据仓库支持百 PB 级数据的离线存储和计算，为业务提供海量、高效、稳定的大数据平台支撑和决策支持。联想在虚拟化和分布式存储领域有超过十年的技术储备，积累了全球 90 多项国际技术专利，有效利用大数据驱动的主动运维和系统优化，为上百家中大型政府企事业机关和企业提供先进云服务。媒体大数据服务商通过人工智能的自然语言处理对互联网媒体大数据进行分析，实现影响力传播监测、评估和决策。

5. 大数据中心建设成为投资热点

全国多个地方加紧布局大数据行业，浙江设立了国家信息经济示范区，贵州、京津冀、珠三角、上海市、河南省、重庆市、沈阳市、内蒙古等地成立国家级大数据综合试验区，积极推进大数据中心建设。2015 年，贵阳 • 贵安大数据产业发展集聚区创建工作获得国家工信部批准，预计到 2017 年年底将建成国家级数据存储灾备基地和国家级云计算应用基地，形成大数据产业集群，产业总规模有望达到 3000 亿元。深圳市将大数据作为重点扶持的领域，吸引了国家超级计算中心（深圳中心）、阿里云深圳数据中心等知名

7 建设成为世界第五大经济体 破解阿里云生态密码，http://www.sohu.com/a/201233452_123988.

大数据企业落户，推动深圳经济创新发展。上海市聚焦“政府治理和公共服务能力提升、经济发展方式转变”两个方面，布局“交易机构+创新基地+产业基金+发展联盟+研究中心”五位一体大数据产业链生态发展，逐步打造国家数据科学中心、亚太数据交换中心和全球“数据经济”中心。

2.4　人工智能技术加速催生新的增长点

1．人工智能技术研发取得诸多成果

2017 年 7 月，国务院印发《新一代人工智能发展规划》，提出面向 2030 年我国新一代人工智能发展的指导思想、战略目标、重点任务和保障措施。目前，南京大学、清华大学、上海交通大学在人工智能核心理论研究方面已取得一系列学术成果，互联网企业在计算机视觉、自然语言处理等人机交互技术方面已经取得一系列研发应用成果。在识别技术研发方面，百度深度学习网络取得人脸识别准确率 99.84%、语音识别准确率 95%等世界领先成绩；腾讯以 83.29%的成绩在国际权威人脸数据库 MegaFace 上 100 万级别人脸识别测试中获得冠军；阿里云 ET 人脸识别技术在户外脸部检测数据库上识别率超过 99.5%。华为成立诺亚方舟实验室专门从事人工智能机器学习、数据挖掘，每年投资超过 500 亿元人民币。科大讯飞是亚太地区最大的智能语音和人工智能上市公司，以语音为切入口，稳步推进从感知智能到认知智能的人工智能生态布局。在人工智能芯片研发方面，2017 年 11 月，中科院孵化的寒武纪科技有限公司发布全球新一代 AI 芯片，包括寒武纪 1H8、1H16、1M 等系列，可用于多种人工智能应用场景。2017 年 6 月，《麻省理工科技评论》发布全球 50 大最聪明企业榜单，中国有 9 家公司入围。

2. 人工智能加速智能机器人研发生产与应用

人工智能技术推动机器人在感知、控制和人机交互能力的大幅提升，机器人通过发挥无人作业、精细操作、便携采集、端云结合等优势，提高工作效率和操作可行性。2015 年，中国工业机器人销量达 6.6 万台，占全球销量总数的 27.02%，但机器人使用密度仅为 49 台/万人，低于全球世界平均水平 69 台/万人[8]，与发达国家相比差距较大。2016 年，中国本土机器人相关企业超过 1 000 家，服务机器人市场规模达到 72.9 亿元，同比大幅增长 44.6%[9]，增长超过工业机器人。随着技术不断赶超以及制造业、服务业转型升级不断深化，我国智能机器人产业市场发展空间巨大。

2.5 新兴前沿技术拓展互联网应用场景

1. 虚拟现实技术发展潜力巨大

虚拟现实技术被《"十三五"信息化规划》、"中国制造 2025"、"互联网+"行动计划等列为重点发展技术，工信部、发改委、科技部、文化部、商务部等出台相关支持政策。各省市地方政府积极建设产业园区及实验室，推动本地虚拟现实产业发展，截至 2016 年年底，有近 20 个省市提出发展虚拟现实产业。目前，我国虚拟现实技术得到广泛应用，主要分为行业应用和大众应用两个方面，行业应用包括城市规划、教育培训、文物保护、医疗、房地产、勘探测绘、生产制造和航空航天等数十个行业；大众应用包括游戏、

8 数据来源：国际机器人联合会。

9 数据来源：易观国际。

社交、影视、直播等领域，“VR/AR+”的时代业已开启。

2. 区块链技术探索和研发正在起步

区块链技术已受到中国各界重视，相关研究相继开展。2016 年 1 月，中国人民银行提出将发行数字货币定为战略目标，加快研究区块链技术在实时交易结算、资源利用和敏感数据保护等金融领域应用问题，并对存在的安全风险和挑战给予密切关注。2016 年 10 月，工信部发布《中国区块链技术和应用发展白皮书（2016)》，首次提出中国区块链标准化路径。目前，中国企业在共识机制、可编程合约、分布式存储、数字签名等核心关键技术方面加强攻关，正在探索建立区块链开源社区、通用开发平台。截至 2017 年 6 月，中国知识产权局数据库已发布 34 个包含“区块链”关键词的有效专利[10]。

10 数据来源：中国知识产权局专利数据库，http://www.pss-system.gov.cn/.

第 3 章　数字经济成为经济增长新引擎

习近平总书记在党的十九大报告中肯定我国“数字经济等新兴产业蓬勃发展”，数字经济已经成为我国以信息化培育新动能、用新动能推动新发展的典型代表。根据二十国集团（G20）杭州峰会发布的《数字经济发展与合作倡议》，数字经济是指以使用数字化的知识和信息作为关键生产要素、以现代信息网络作为重要载体、以信息通信技术的有效使用作为效率提升和经济结构优化的重要推动力的一系列经济活动。当前，我国积极推动数字产业化和产业数字化，在电子商务、信息服务、网络终端设备、互联网金融、共享经济及“互联网+”融合等领域取得了突出成效。2016 年我国数字经济规模达到 22.58 万亿元人民币，总量跃居全球第二，增速高达 21.2%，占 GDP 比重达到 30.3%[1]，成为经济增长的新引擎和新亮点。

3.1　网络终端设备制造加快升级

3.1.1　中国智能手机逐鹿全球市场

1. 国内智能手机市场进入稳定期

我国企业从 2010 年开始布局智能手机业务发展，7 年来增长迅速，国

1 数据来源：中国信息通信研究院，《中国数字经济发展白皮书（2017 年）》。

产手机已经占领国内智能手机市场。2017 年上半年国内智能手机销售量前四名均为本土厂商，其中华为以 21.6%的销售占比排名居首，OPPO、vivo、小米分别列第二、三、四位[2]，三星、苹果等国外品牌手机占有率持续下降。国内智能手机市场进入稳定期，2016 年保有量为 10.6 亿台，增长率持续下降，未来市场逐步饱和，增长幅度趋于平稳。2011—2017 年我国智能手机保有量如图 3-1 所示。

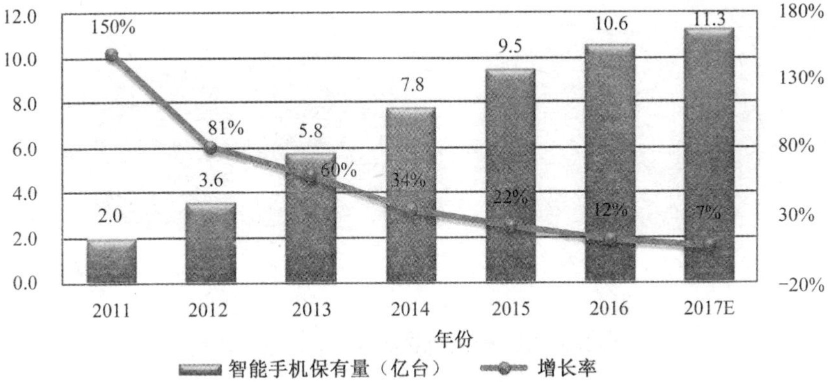

图 3-1　2011—2017 年我国智能手机保有量

（数据来源：中国产业发展研究网）

2. 国产智能手机抢占全球市场

近年来，国产智能手机行业经过激烈的竞争和淘汰迅速崛起，国际竞争力和影响力显著提升，国际市场份额持续增加，成为我国高端制造走向国际化的一个缩影。2016 年，中国智能手机全球出货量 4.65 亿台，约占全球 15

2 数据来源：艾媒咨询，《2017 上半年智能手机市场研究报告》。

亿台出货量的$\frac{1}{3}$[3]。国产领军品牌已和国际知名品牌平分秋色，海外市场开拓初见成效。中兴智能手机覆盖全球 160 个国家和地区，与 230 家运营商建立合作关系；华为基于海思芯片，推动智能手机向高端化发展，2016 年出货量增幅达到 21%，远高于三星和苹果手机的增幅；OPPO、vivo 瞄准都市时尚群体，2016 年全球出货量增幅分别高达 109%和 78%。2012 年和 2017 年国内智能手机品牌全球市场占有率对比如图 3-2 所示。

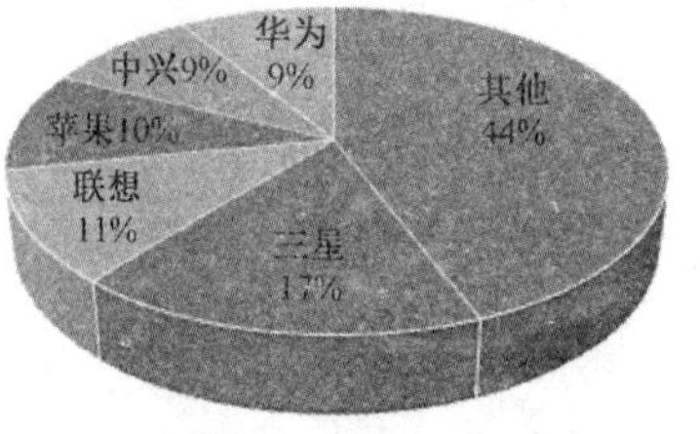

（a）2012年全球智能手机品牌市场份额

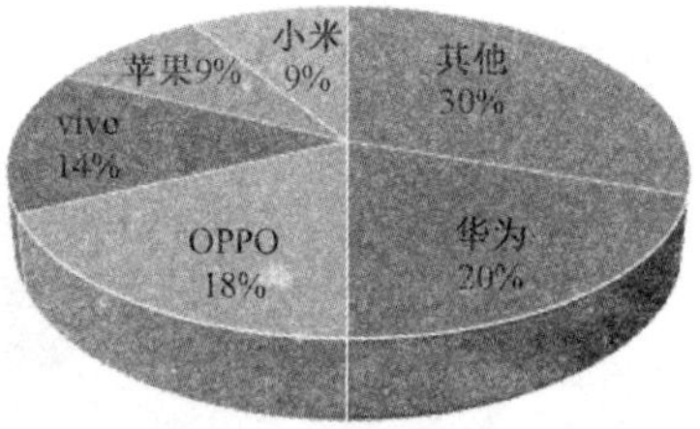

（b）2017年第一季度全球智能手机品牌市场份额

图 3-2　2012 年和 2017 年国内智能手机品牌全球市场占有率对比

（资料来源：新浪财经）

3.1.2　新型智能终端蓬勃发展

新技术的快速发展，助推智能家居、车联网、人工智能、虚拟现实等领域智能终端迅速增长，产品快速迭代。

1. 智能家居生态逐步建立

中国智能家居行业近年发展迅速，吸引了包括传统硬件企业、互联网企业、房地产企业、家装企业等纷纷抢滩，百度、华为、小米、联想等科技公

3 数据来源：Counterpoint Research，《2016 全球手机市场调研报告》。

司加大投资力度，推动人工智能、语音识别、深度学习等技术与智能家居技术深度融合。随着 5G、蓝牙 5.0、下一代 Wi-Fi 等移动通信技术的商业化应用，智能家居行业将迎来新的变革，产品更加丰富多样，产业潜力加速释放，产业生态逐步建立。智能电视作为典型代表领域，市场规模突飞猛进。2016 年，中国智能电视销量达 4 098 万台，预计到 2018 年将突破 5 000 万台（见图 3-3）。

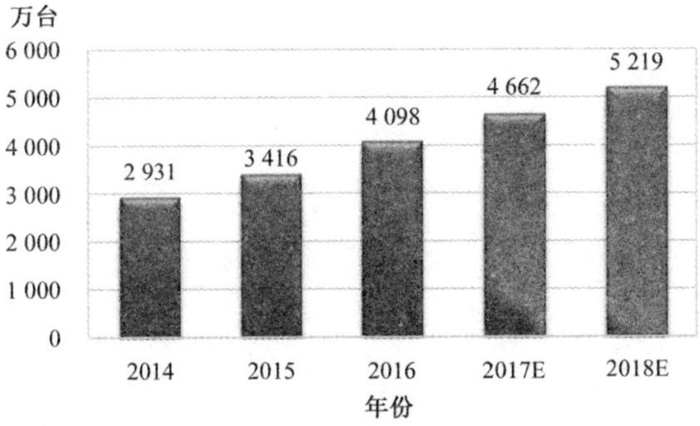

E 指预测

图 3-3　2014—2018 年中国智能电视销售量及预测

（数据来源：艾媒咨询，《2016—2017 中国智能电视市场专题研究报告》）

2．车联网终端快速发展

互联网企业、芯片厂商等将车载操作交互系统和车载处理器作为新突破口，纷纷进入车联网市场。传统汽车制造商积极推进车联网战略，宝马自 2015 年 6 月起所有在华销售车型实现 100%联网，上海通用汽车计划于 2020 年实现旗下产品 100%互联。我国主导的车联网 LTE-V2X 标准已基本成熟，2017 年 3 月在国际标准组织 3GPP 完成标准化工作。截至 2017 年 8 月，中国移动、中国联通、中国电信三大运营商的车联网用户数合计突破 5 806 万户，共 683 款车载联网终端获工信部入网证。其中，4G 车载联网终端有 101 款[4]。

4 数据来源：中国信息通信研究院，《车联网白皮书（2017）》。

3．人工智能产业爆发增长

中国人工智能产业起步相对较晚，但随着科技、制造等业界巨头公司的布局深入，以及众多垂直领域创业公司的诞生和成长，越来越多的产业级和消费级应用产品正在涌现。2016 年我国人工智能应用创新集中爆发，当年 3 月中科院发布全球首款神经网络处理器“寒武纪”，7 月人工智能金融服务商第四范式发布开发平台“先知”，8 月阿里巴巴发布人工智能 ET，9 月百度对外展示“百度大脑”，10 月蚂蚁金服联合清华大学成立 AI 实验室，12 月腾讯云向全球发布 7 项 AI 服务，在各行业细分领域直追国际领先水平。2016 年，中国人工智能产业市场规模已突破 100 亿元，相比上年增长 43.3%，预计 2018 年规模将达 238 亿元（见图 3-4）。

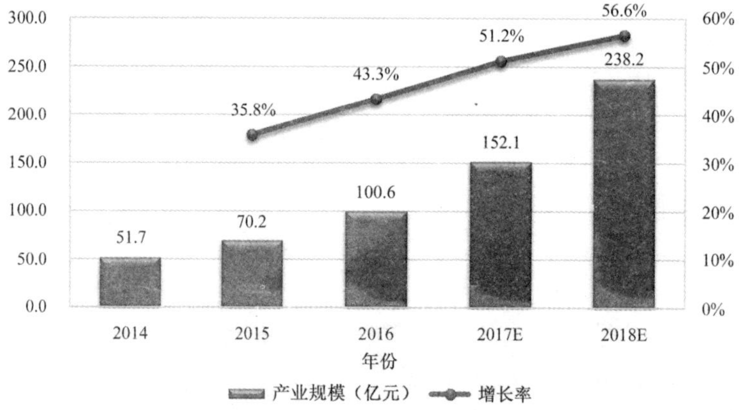

图 3-4　2014—2018 年中国人工智能产业规模及预测

（数据来源：艾媒咨询，《2017 年中国人工智能产业专题研究报告》）

4．虚拟现实应用快速推进

我国虚拟现实应用市场总体规模不大，但增速迅猛，主要代表厂商有维阿时代、暴风魔镜、乐蜗科技、凌感科技、诺亦腾等。2016 年中国 VR 市场规模为 34.6 亿元，预计未来五年 VR 市场年复合增长率将超过 80%，2018 年整体规模将突破百亿元大关，2021 年规模达到 790 亿元，有望成为全球最大的 VR 市场。其中，规模最大的细分领域是 VR 头戴设备，2016 年以 20.5 亿元的规模占据整体份额的 59.2%；VR 体验馆紧随其后，预计市场规模将在 2021 年达到 52 亿元；运动捕捉系统、触感控制器、声音采集设备等其他 VR 硬件市场，预计将以年复合增长率 54.6%的速度增长，2021 年将达到 36 亿元的规模[5]（见图 3-5）。

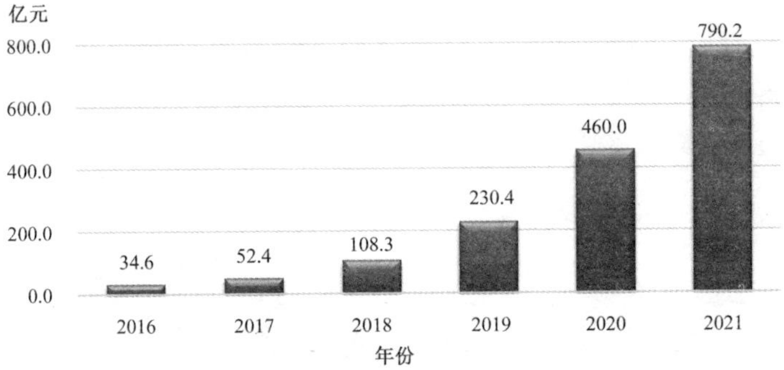

图 3-5　2016—2021 年中国 VR 市场规模及预测

（数据来源：艾瑞咨询&Greenlight Insights《中国虚拟现实 VR 行业研究报告》）

3.1.3　网络设备可信体系逐步建立

1. 自主芯片产业发展势头良好

“十二五”期间，我国集成电路扶持政策力度不断加强，特别是国家集成

5 数据来源：艾瑞咨询联合 GreenlightInsights 发布的 2017 年《中国虚拟现实 VR 行业研究报告——市场数据篇》。

电路产业发展投资基金的设立，推动自主芯片产业快速发展，成为全球增长最快的地区之一，2013—2016 年产业销售收入年均增速 20%，2016 年产业规模达到 4 336 亿元[6]。中国内地芯片产能全球占比达到 10.8%，已建成 9 条 12 英寸晶圆线、另有 8 条在建[7]。我国龙头企业持续加强研发强度，海思、展讯等移动端处理器芯片已能满足使用要求，同方国芯、紫光国芯、武汉新芯、福建晋华等企业纷纷投资新建芯片工厂，收购芯片产业链上下游企业。中国电子等企业围绕芯片构建产业生态，形成了从芯片到硬件整机，再到终端设备、传感、通信网络、操作系统、平台及应用的完整布局，未来我国在全球芯片市场的竞争力将进一步提升。2009—2016 年中国集成电路产业销售规模如图 3-6 所示。

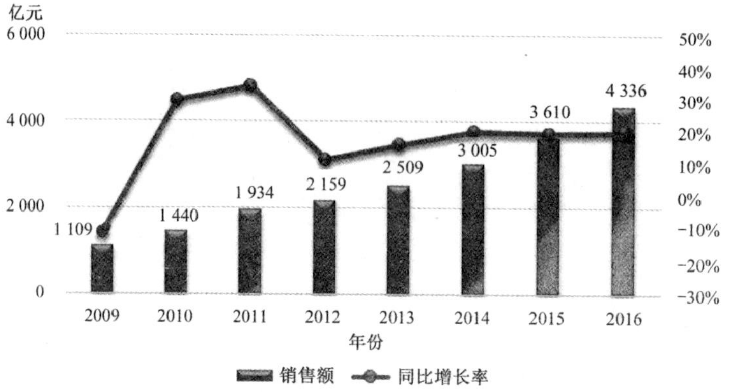

图 3-6　2009—2016 年中国集成电路产业销售规模

（数据来源：中国半导体行业协会）

2．传感器产业发展稳定

我国从“七五”期间就开始发展传感器产业，经过 30 年的稳定发展，

6 数据来源：中国半导体行业协会，中国半导体行业信息网（www.csia.net.cn）。

7 数据来源：IC Insight，《全球晶圆产能预测报告 2017—2021》。

已形成完整产业链。目前，正积极突破敏感元件核心技术及生产工艺，加快部署应用到工业生产、医学诊断、海洋探测、环境保护等领域，朝着高端化、智能化、微型化方向发展。2016 年，中国传感器市场规模达到 1 126 亿元（见图 3-7），同比增长 13.2%，继续保持稳步增长。

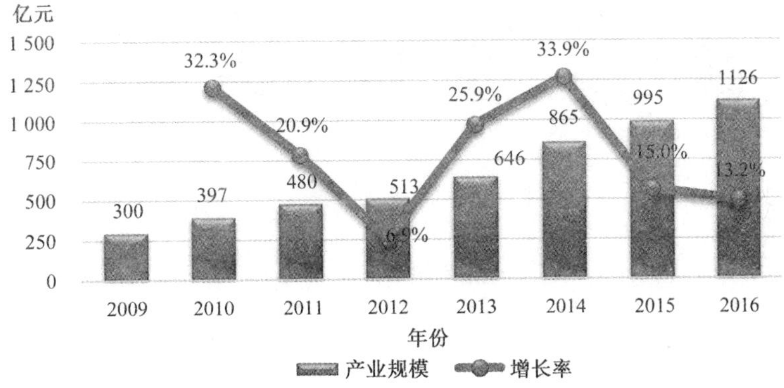

图 3-7　2009—2016 年中国传感器制造行业市场规模

（数据来源：智研咨询，《2017—2023 年中国传感器制造行业运营态势与投资战略咨询报告》）

3．计算机设备保持市场优势

近年来，由于国内外市场需求的持续疲软与移动互联网的进一步兴起，微型计算机产品产量持续下滑；联想保持霸主地位，从 2013 年开始连续 4 年成为全球最大的个人计算机生产厂商[8]。服务器、存储系统等领域维持增长态势。2016 年，全球服务器出货量排名前五的企业中，我国企业占据三席，分别为华为、联想、浪潮。此外，以“神威·太湖之光”为代表的超级计算机、以“浪潮天梭 K1”为代表高端容错计算机陆续面世，推动我国高端计算机行业加快发展。2011—2016 年中国微型计算机产品产量与增长率

8 数据来源：Gartner 历年全球 PC 市场销量报告。

如图 3-8 所示，2016 年全球排名前 5 名的服务器厂商出货量如图 3-9 所示。

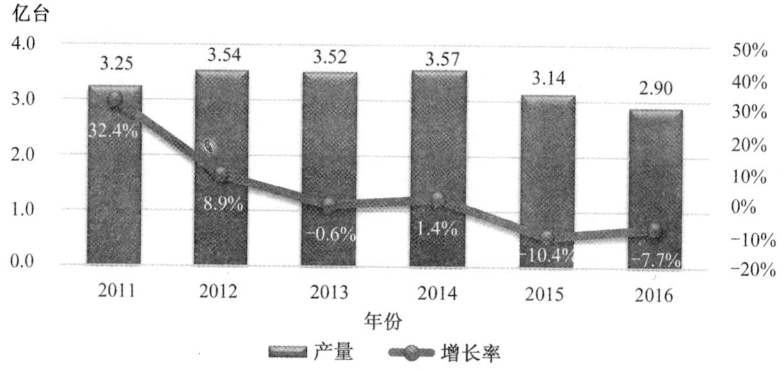

图 3-8　2011—2016 年中国微型计算机产品产量与增长率

（数据来源：赛迪智库，2017 年 2 月）

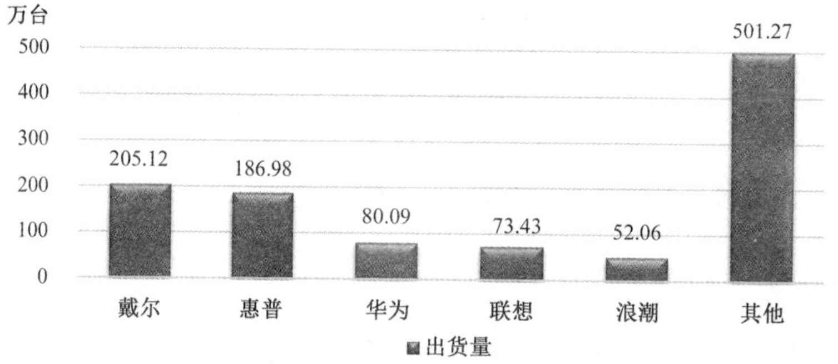

图 3-9　2016 年全球排名前 5 名的服务器厂商出货量

（数据来源：Gartner 2016 年第 1～4 季度市场监测数据）

4．安防设备产业异军突起

经过 30 年的发展，我国安防设备产业逐步从模拟化、数字化走向智能化。在国内市场，随着平安城市、民用安防等市场需求的逐步升温，以视频监控为主体的智能安防设备开始占据基础设施建设的重要位置。2016 年，我国安防行业市场规模达到 5 400 亿元（见图 3-10），自 2012 年以来年均复合增速达到 15%。在 G20 峰会、里约奥运会等安保服务中的大展身手，让全世界看到了中国安防产品和技术的领先水平。2016 年全球安防 50 强企业中，海康威视位列第一，大华技术位列第四；其中，在视频监控领域，海康威视已经连续六年蝉联全球市场占有率第一位[9]。

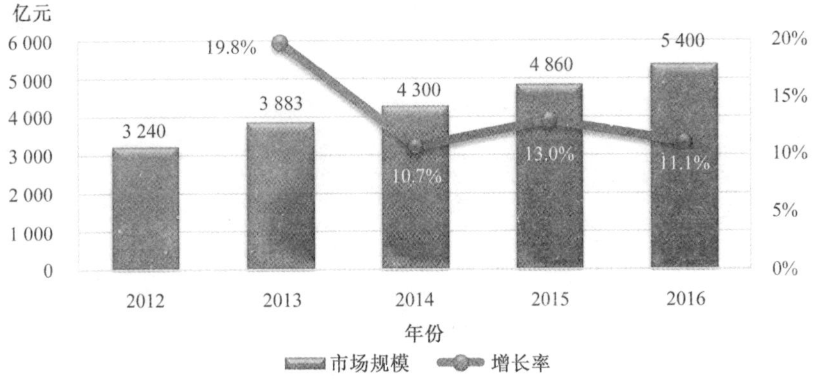

图 3-10　2012—2016 年中国安防行业市场规模

（数据来源：智研咨询，《2017—2022 年中国安防行业深度评估与投资前景研究报告》）

3.2　互联网信息服务产业化进程加速

随着“宽带中国”战略、4G 网络建设等持续推进，我国信息消费呈现

9 数据来源：国际市场调研机构 HIS 发布数据。

出欣欣向荣的局面，消费规模由 2013 年的 2.2 万亿元增长至 2016 年的 3.9 万亿元，年均增幅 21%，为同期传统消费增速的 2.4 倍，对 GDP 增长贡献率达到 0.26 个百分点[10]。随着国务院《关于进一步扩大和升级信息消费 持续释放内需潜力的指导意见》的出台，互联网信息服务将迎来新一轮的高速增长期。

3.2.1 网络内容产业繁荣发展

1. 网络娱乐产业快速兴起

网络视频、网络音乐、网络文学、网络游戏等网络娱乐产业全面迅速崛起，用户规模显著增长，成为文化产业的重要组成部分。截至 2017 年 6 月，中国网络视频用户规模为 5.64 亿人（见图 3-11），网络音乐用户规模为 5.24 亿人（见图 3-12），网络文学用户规模为 3.52 亿人（见图 3-13），网络游戏用户规模为 4.21 亿人（见图 3-14）。2016 年，网络娱乐产业市场规模达 2 631.46 亿元，其中，网络游戏市场 1 789.2 亿元，同比增长 24.6%[11]（见图 3-15）；网络视频市场 609 亿元，同比增长 55.9%[12]；网络音乐市场 143.26 亿元，同比增长 39.4%[13]；网络文学市场约 90 亿元，同比增长 28.6%[14]。网络娱乐行业之间正在加速融合发展，以腾讯、网易、巨人网络为代表的多家

10 数据来源：中国新闻网记者 唐贵江，中国工信部副部长：去年中国信息消费规模达 3.9 万亿元，中新社广州 2017 年 3 月 23 日电。

11 数据来源：艾瑞咨询，《2016 年中国网络游戏行业研究报告》。

12 数据来源：国家新闻出版广电总局发展研究中心和网络视听节目管理司，《中国视听新媒体发展报告（2017）》。

13 数据来源：中国传媒大学音乐与录音艺术学院，《中国数字音乐产业发展报告》。

14 数据来源：产业研究智库，《2016 年中国网络文学行业规模现状及态势分析》。

厂商以知识产权创意为核心，积极布局视频、直播、游戏、文学、社交、电影票务等泛娱乐业态，实现集聚、融合、联动发展。

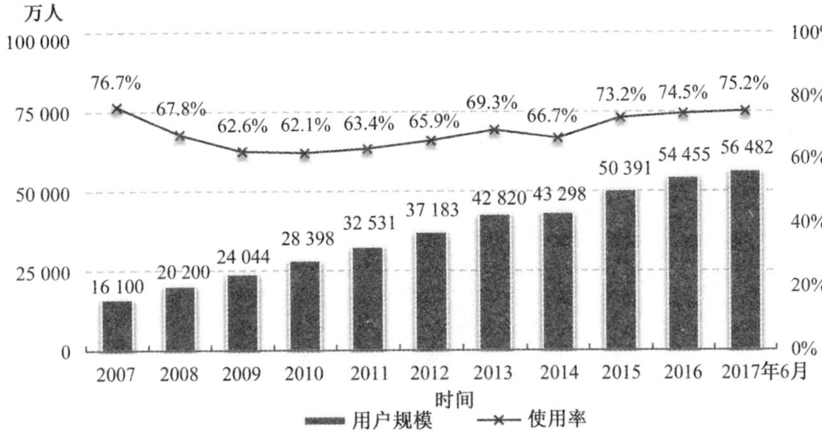

图 3-11 2007—2017 年 6 月中国网络视频用户规模和使用率

（数据来源：CNNIC，第 39、40 次《中国互联网络发展状况统计报告》）

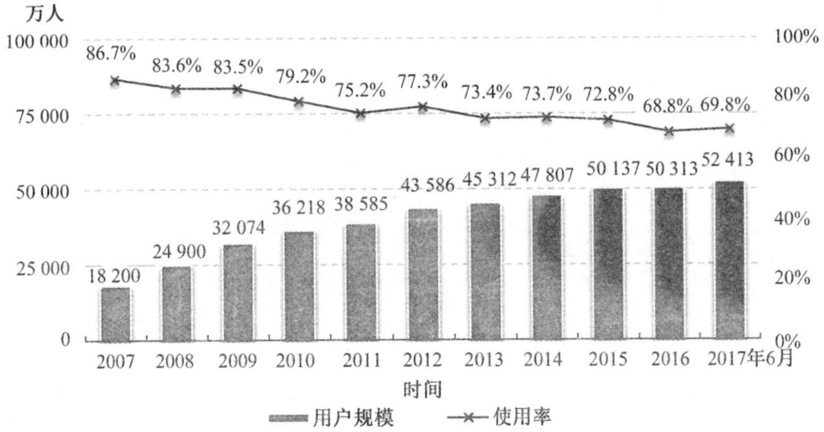

图 3-12 2007—2017 年 6 月中国网络音乐用户规模和使用率

（数据来源：CNNIC，第 39、40 次《中国互联网络发展状况统计报告》）

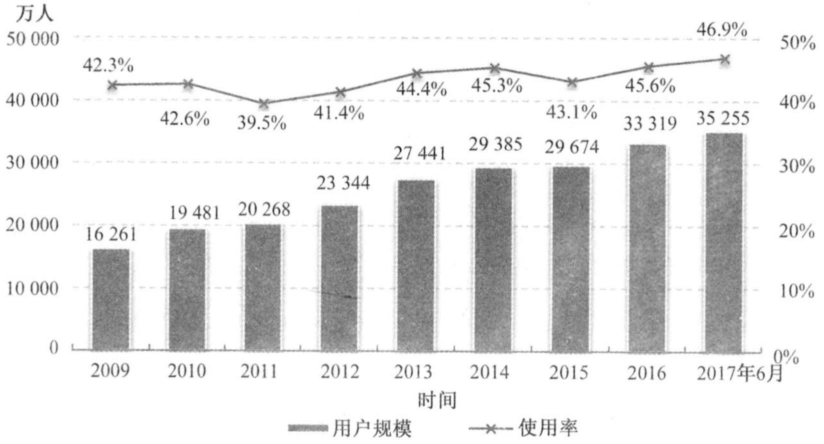

图 3-13　2009—2017 年 6 月中国网络文学用户规模和使用率

（数据来源：CNNIC，第 39、40 次《中国互联网络发展状况统计报告》）

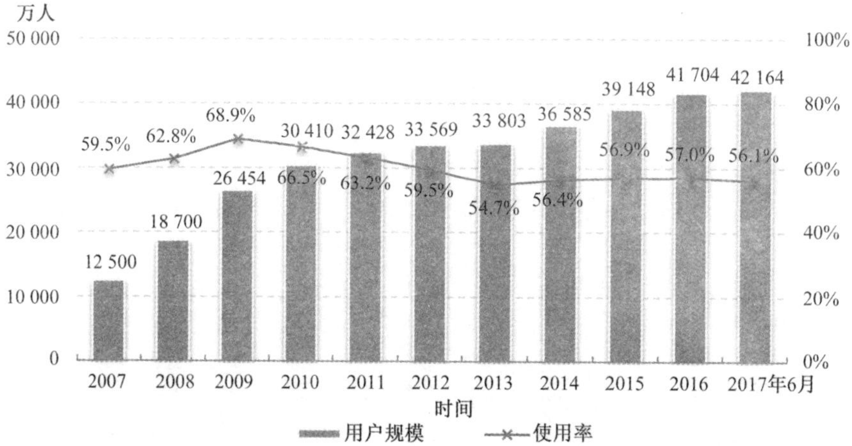

图 3-14　2007—2017 年 6 月中国网络游戏用户规模和使用率

（数据来源：CNNIC，第 39、40 次《中国互联网络发展状况统计报告》）

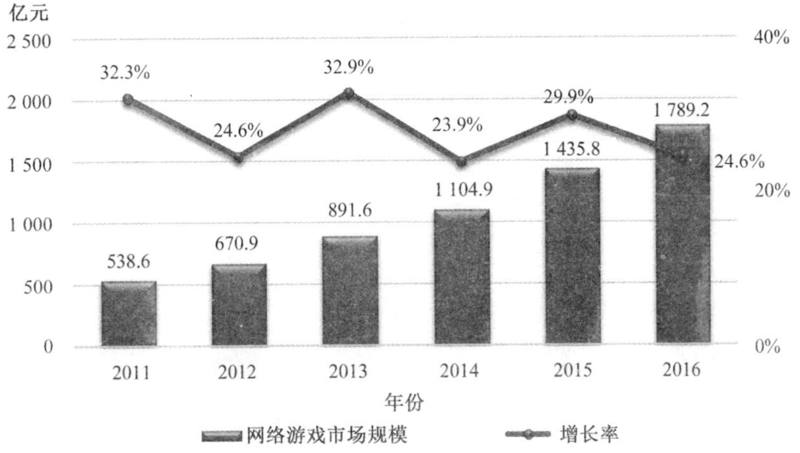

图 3-15　2011—2016 年中国网络游戏市场规模及增长率

（数据来源：艾瑞咨询，《2016 年中国网络游戏行业研究报告》）

2. 网络广告占比不断扩大

自 2012 年“程序化购买元年”起，受网民人数、数字媒体使用时长、网络视听业务等因素快速增长推动，网络广告市场份额快速增长，而电视、广播、报纸、杂志等传统广告市场份额持续下滑。2016 年，中国网络广告市场规模达 2 902.7 亿元，同比增长 32.9%，在五大媒体广告收入中的占比已经达到 68%。其中，移动广告市场规模为 1 750.2 亿元，同比增长率高达 75.4%（见图 3-16），占比首次过半，达到 60.3%。目前，我国网络广告的形式不断创新，大数据应用及分析能力不断提升，社交、新闻资讯、视频网站中的信息流广告快速成长。同时，网络直播平台及网红概念的兴起也不断影响网络广告的形式与创意，内生广告可能成为未来的发展趋势。

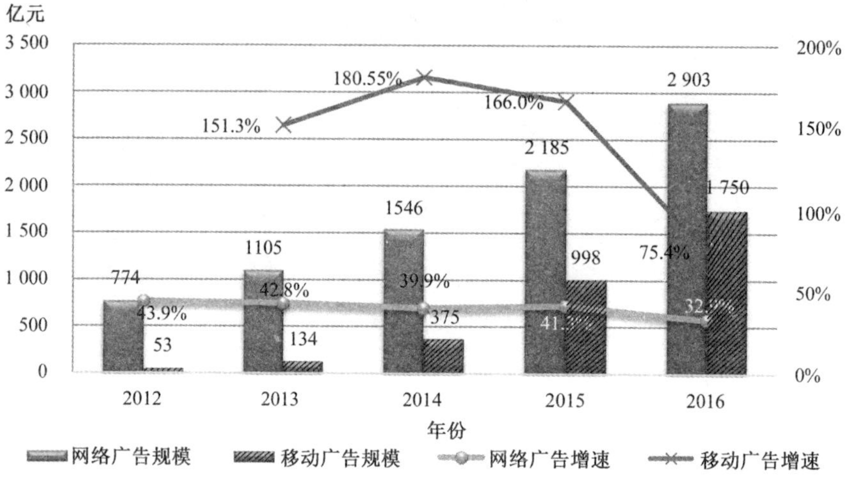

图 3-16　2012—2016 年中国网络广告和移动广告市场规模及增速

（数据来源：艾瑞咨询，《2017 年中国网络广告市场年度监测报告》）

专栏 4　网络广告的程序化购买

程序化购买就是基于自动化系统（技术）和数据来进行的广告投放。它旨在通过利用客户的数据和洞察，在合适的时间，合适的环境中覆盖合适的用户来提高数字广告投放的效率与效果。

自 2012 年开始，程序化购买受到资本的热烈追捧，上百家程序化购买平台先后出现，经过五年的发展，已经形成媒体方、广告交易市场、程序化买家、监测方、广告代理公司、广告主甲方共同参与的市场体系，流量规模及质量、技术基础设施、交易模式、各方意识等各方面已日趋成熟。然而，程序化购买在发挥技术和数据在广告投放中重要作用的同时，也出现了流量作弊、虚假交易、投放过程不透明、品牌安全受到伤害等问题。这些问题经过多年累积，在 2016 年集中爆发，程序化购买受到广告主的质疑，并伴随交易规模增速的下滑，网络广告程序化购买市场迎来调整期[15]。

3.2.2 搜索引擎市场持续扩大

搜索引擎发挥着连接用户与服务环节的桥梁作用。近年来，我国搜索引擎市场稳步增长。2016 年，中国搜索引擎市场收入为 805 亿元，增长率为 4.5%，占全球搜索引擎市场总收入超过 12.5%，预计未来收入规模存在更大发展空间[16]。截至 2017 年 6 月，我国搜索引擎用户规模增长至 6.09 亿人，使用率为 81.1%（见图 3-17）。

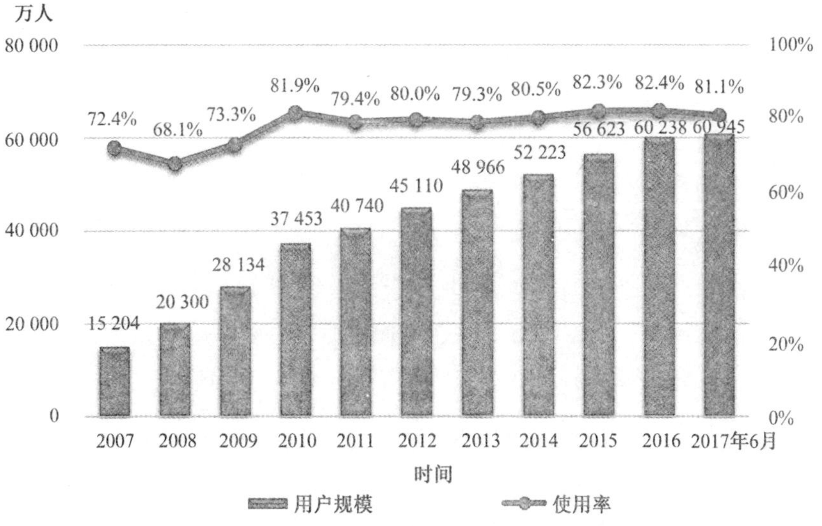

图 3-17 2007—2017 年 6 月中国搜索引擎用户规模和使用率

（数据来源：CNNIC，第 39、40 次《中国互联网络发展状况统计报告》）

15 艾瑞咨询，《2017 年中国网络广告市场年度监测报告》。

16 数据来源：易观产业数据库，《中国搜索引擎市场季度监测报告》（2016 年第 1～4 季度）。

1. 搜索技术创新不断

随着新兴信息技术的不断涌现，百度、搜狗等主流搜索引擎不断加大在大数据、人工智能等前沿技术方面的投入，大幅提升用户搜索体验。目前，市场上主流搜索引擎的语音、图像、视频等识别技术已经能够达到较高成功率，进一步实现“所想即所搜、所搜即所得”。在语音识别方面，手机百度的语音搜索相对错误率降低了25%以上，中文语音识别率突破90%；在图像搜索方面，百度图片搜索的准确率从原来的20%飙升到80%，评测效果超过脸谱（Facebook）的搜索技术，视觉搜索平台的日检索量超过5 000万次。同时，人工智能机器人辅助搜索已经成为各大搜索引擎的标准配置，如百度的“度秘”、搜狗的“语音助手”、微软必应的“小冰”等，受到用户广泛认可。

2. 搜索引擎服务模式加快转型发展

搜索引擎行业垂直化、专业化发展趋势明显，搜索服务商业模式趋于多元，盈利模式不再限于单纯的广告营销收入，搜索模式从单一的信息搜索服务向网络综合服务平台转型，各大搜索平台应用大数据等新技术打通地图、购物、生活、新闻、医疗等多种内容的搜索服务，满足用户在新闻热点、商品消费等不同领域的多样化搜索需求。例如，搜狗搜索相继接入微信、QQ兴趣部落、知乎等，并与微软必应达成合作，在社交、新闻、专业问答、英文和学术搜索等垂直领域强化优质内容的吸收力度，构建新型内容生态；百度响应连接消费支付场景潮流，推出搜索直达的实物商品、本地生活服务、金融产品、文化产品等服务；360 搜索、神马搜索等也纷纷上线网络购物和O2O 生活服务平台业务。

3. 移动搜索成为市场发展主推动力

自 2011 年以来，手机搜索用户规模年均增速一直快于整体搜索用户规模增速，截至 2017 年 6 月，手机搜索用户数达 5.93 亿人（见图 3-18），比 2016 年年底增加 1 760 万人，增长率为 3.1%，比整体搜索用户增长率高 1.9 个百分点。2017 年移动端搜索使用率为 81.9%，全面超越 PC 端，移动营收在整体营收增长中的贡献也越来越大。2016 年，百度搜索全年移动营收在总营收中占比达到 63%；搜狗搜索移动端流量增长 70%，对整体流量的贡献达到 75%[17]。

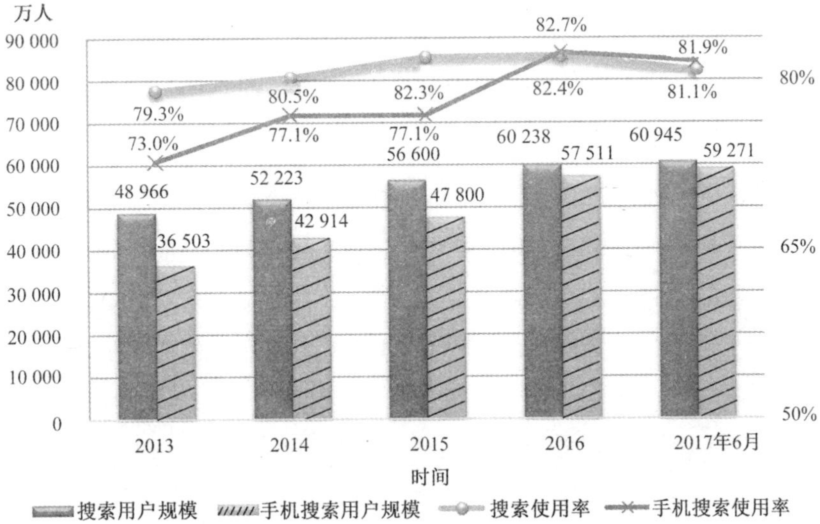

图 3-18　2013—2017 年 6 月手机搜索用户规模和搜索使用率对比

（数据来源：CNNIC，第 39、40 次《中国互联网络发展状况统计报告》）

17 数据来源：百度 2016 全年财报、搜狗 2016 全年财报。

3.3 电子商务发展势头强劲

电子商务作为一种新型交易方式，将生产企业、流通企业及消费者和政府带入了数字化商业新纪元。我国政府大力支持电子商务发展，先后发布《关于加快电子商务发展的若干意见》《电子商务发展“十一五”规划》《关于大力发展电子商务加快培育经济新动力的意见》等政策文件，明确电子商务发展的战略、目标和举措，不断完善政策环境、监管环境、市场环境、金融环境、物流环境等，推动中国电子商务实现了长达二十年的快速发展，成为当前中国数字经济最为活跃和最具代表性的行业。1997 年，我国首家企业对企业（B2B）电商平台——“中国化工信息网”正式上线，一年后，阿里巴巴、中国制造网等B2B企业相继成立；1999 年，第一家消费者对消费者（C2C）电子商务平台——易趣网、第一家企业对消费者（B2C）电子商务平台——8848 成立；2003 年，淘宝网、京东商城等一批 B2C 平台相继崛起[18]，推动网络零售交易额持续高速增长，带动移动电商、农村电商、跨境电商等新领域蓬勃发展、蒸蒸日上。2016 年，我国网络购物用户人数增长至 4.67 亿人，电子商务交易额增长至 26.1 万亿元（见图 3-19），相当于国民生产总值的 35%，自 2004 年以来保持高达 31.24%的年均复合增长率，直接和间接带动就业人数 3 700 万人，创造的年度税收超过 2 000 亿元，成为全球电子商务最为发达的国家之一。

18 王炳南：《第十二届全国人大常委会专题讲座第二十九讲：国内外电子商务的现状与发展》，2017 年 6 月 29 日。

图 3-19　2004—2016 年中国电子商务交易总额及增长率

（数据来源：历年商务部《中国电子商务报告》）

3.3.1　B2C 与移动端成为网络零售趋势

1. 网络零售全球居首，品类模式不断丰富

网络零售是我国电子商务的重要组成部分与发展亮点。自 2008 年以来，我国网络零售额[19]以年均 32%以上的复合增长率迅速扩大。2013 年，中国网络零售交易额达 1.9 万亿元，首次超过美国，成为世界第一大网络零售市场。2016 年中国网络零售交易额达 5.16 万亿元（见图 3-20），同比增长 26.2%，对社会消费品零售总额增加值的贡献率增长至 30%；其中实物商品的网络零售额为 4.19 万亿元，比同期社会消费品零售额增速高出 15.2%。同时，网络零售的用户、商品、模式基础不断稳固，网络零售商品品类不断丰富，网络零售市场新业态新模式频现，“新零售”引爆线上线下融合发展，不断满足用户日益增长的购物需求。

19　网络零售额是指通过自建网站和第三方平台等公共网络交易平台实现的商品和服务零售额。

图 3-20　2008—2016 年中国网络零售交易额及增长率

（数据来源：历年商务部《中国电子商务报告》）

专栏 5　“双十一”

“双十一购物狂欢节”是指每年 11 月 11 日的网络促销日，最初源于淘宝商城（天猫）2009 年 11 月 11 日举办的促销活动，尽管当时参与的商家数量和促销力度相对有限，但营业额远超预想的效果，使得 11 月 11 日成为天猫举办大规模促销活动的固定日期，并且逐渐蔓延到其他电商平台、厂家和运营商，壮大成为中国电商行业的年度盛事。

2014 年 11 月 11 日，阿里巴巴“双十一”全天交易额达到 571 亿元；2015 年 11 月 11 日，天猫双十一全天交易额为 912.17 亿元；2016 年 11 月 11 日，这一数字更是突破 1 207 亿元，其中无线交易额占比高达 81.87%。在 11 月 11 日至 16 日的 6 天时间里，全国邮政、快递企业共揽收快件 11.2 亿件，派送 7.9 亿件。2017 年 11 月 11 日，阿里巴巴全天交易额达到 1 682 亿元，京东商城累计下单金额超过 1271 亿元。“双十一”已经成为世界性的电商狂欢节，北到格林列岛、南到智利，网购活动覆盖 235 个国家和地区。"双十一"订单量和交易额的连年爆发式增长，表明线上交易形式已经由之前的作为零售产业的补充渠道之一，转型为拉动中国内需的主流形式，由此开始全面倒逼传统零售业态升级。

2. B2C 成为网络零售主流，市场份额超过 C2C

中国电子商务发展初期，C2C 发展较快、规模较大。随着消费能力升级，B2C 逐步替代 C2C 成为主流。2015 年，B2C 市场交易规模达到 2 万亿元，占比 51.9%，首超 C2C 份额[20]。2016 年，B2C 网购市场交易规模达到 2.6 万亿元，较 2015 年增长 31.6%，远远超过 C2C 市场 15.6%的增速[21]。随着用户消费需求进一步转变，C2C 平台交易规模占比将逐年下滑，B2C 市场的模式创新层出不穷，反向 C2B 定制、宅配、众筹、社交拼团、一元购、品质优选、原产地直供、周期购等新模式陆续面世，推动 B2C 网络零售市场持续增长。

专栏 6　阿里巴巴的平台电商模式

阿里巴巴集团于 1999 年在杭州创立，目前已经发展成为拥有完整互联网商务生态系统的大型集团，主营和关联业务包括：淘宝网、天猫、聚划算、全球速卖通、阿里巴巴国际交易市场、1688、阿里妈妈、阿里云、蚂蚁金服、菜鸟网络等。2014 年 9 月 19 日，集团在纽约证券交易所正式挂牌上市；2016 年 4 月 6 日，阿里巴巴正式宣布已经成为全球最大的零售交易平台。

阿里巴巴旗下的"淘宝"和"天猫"是典型的平台型电商，平台的定位是提供展示机会和流量来源，并且为互联网支付提供信用保障和各类服务接口；而产品的设计、研发、制造、营销、仓储、配送和售后仍然由卖家完成，卖家以市场需求为导向，形成品类繁荣的大集市。阿里

20 数据来源：艾瑞咨询，《2015 年中国电子商务数据报告》。

21 数据来源：艾瑞咨询，《2016 年中国电商生命力报告》。

巴巴的电商平台专注于信息流，资金流和物流则通过蚂蚁金服、菜鸟网络进行整合；在汇聚大量的市场供求信息的基础上，通过增值服务为会员提供优越的市场服务。阿里巴巴以“让天下没有难做的生意”为使命，始终保持着平台的开放性，在集聚大量商家、用户的同时，也将生态系统开放给互联网服务的创业者，从而不断地衍生出新业务模式，形成巨大的产业链闭环。

专栏 7　京东的自营电商模式

京东于 2004 年正式涉足电商领域。2014 年 5 月，京东成为中国第一个成功赴美上市（纳斯达克）的综合型电商平台。十三年间，京东交易额增长 9 万倍，年均复合增长率超过 150%，服务了中国 9 亿多用户和十几万品牌商，拥有超 12 万名正式员工并间接创造 400 万个就业机会[22]。

京东商城致力于打造一站式综合购物平台，依托电商、金融、物流三大业务板块，创造了自营电商的发展模式。与京东合作的厂家，专注在产品的研发生产阶段，流通环节全部交由京东处理；而京东以交易为基础，延伸至仓储、配送、售后、营销四个流通环节，提供互联网零售领域的完整服务。

京东的电商模式取得了长足发展，其家电销售份额占据国内家电网购市场份额的 62%，3C 类占据线上市场份额的 50%；京东时尚成为增速最快的品类，在新用户购买品类中占据 40%以上；京东“百万便利店计划”服务于全国 680 万家中小门店；“京东企业购”占据企业级市场份额的 45%以上。

3. 网购进入移动消费时代

近年来，中国的移动购物发展迅猛，成为网络零售的新增长点。2015

22 数据来源：京东 2016 年财报。

年，移动端网购使用率达 54.8%，首次超过 PC 端，成为网络购物的主渠道。2016 年，移动端消费持续扩大，用户规模迅速增长到 4.69 亿人，使用率提高到 67.5%（见图 3-21），移动购物市场规模增长至 3.3 万亿元，同比增长 57.9%，占网络购物总交易规模的 70.2%[23]。移动端已成为各大电商企业竞相发力的主战场，以天猫为代表的传统成熟电商企业、以蘑菇街等为代表的新兴移动电商企业均纷纷在移动端布局，打造立体化生态格局。企业营销方式加速多元化，90%的电商企业推出移动应用 APP[24]，加大移动端数字营销广告投入，加快推广服务消费模式，引入媒体元素进行兴趣导购，探索视频电商导购模式等。2016 年，我国网购市场排名前十的企业移动端增速达到 27.1%，是 PC 端增速的 3 倍[25]。

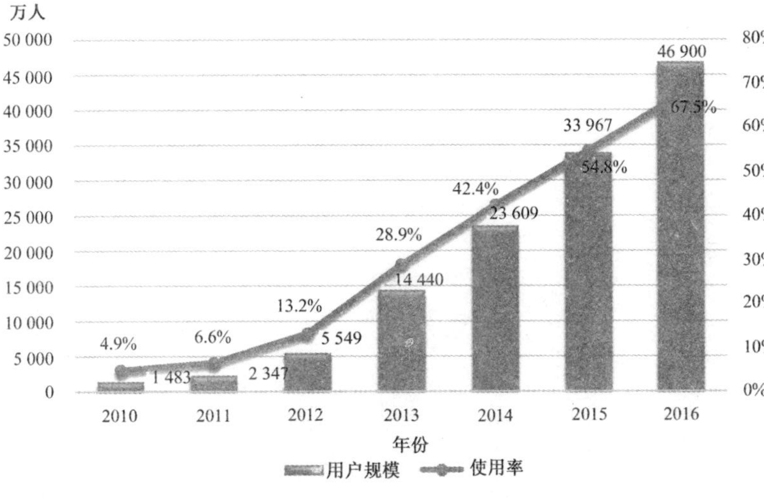

图 3-21　2010—2016 年中国手机网购用户规模和使用率

（数据来源：CNNIC，第 39 次《中国互联网络发展状况统计报告》）

23 数据来源：艾瑞咨询，《2017 年中国移动电商行业研究报告》。

24 中国电子信息产业发展研究院，《2016—2017 年中国工业和信息化发展系列蓝皮书》。

25 数据来源：艾瑞咨询，《2017 年中国移动电商行业研究报告》。

3.3.2 垂直电商平台发展加速

1. B2C 类垂直电商平台加快精细化发展

B2C 类垂直电商瞄准各自细分领域，围绕产品精细化、服务专业化、营销社会化等方面精耕细作、精准发力，满足消费者多元化、个性化需求。在美妆、服装等时尚垂直电商领域，有唯品会、聚美优品、蘑菇街、乐蜂网等多家电商，其中唯品会、聚美优品在美国纽交所上市，成为行业竞争的佼佼者。食品类垂直电商领域，我买网依托中粮集团的产品供应体系、顺丰优选依托顺丰的生鲜物流配送优势，在众多同行中树立起独特的竞争优势。酒类电商酒仙网、母婴类电商红孩子、宠物类电商波奇网、珠宝类电商东方美宝、箱包类电商麦包包，以及汽车后市场类电商途虎养车、汽车超人、瓜子二手车等都在各自细分领域占据了一席之地。

专栏 8 唯品会的时尚特卖模式

唯品会是中国垂直电商平台的领军企业之一，开创了"精选品牌、深度折扣、限时抢购"的时尚特卖模式。2008 年 8 月，唯品会成立于广州；2012 年 3 月，唯品会登陆美国纽约证券交易所。2016 年唯品会净营收 565.9 亿元，年订单量 2.698 亿单，日均订单量达 80 万单[26]。

唯品会主要销售的品类包括服饰鞋包、美妆、家居生活等，累计合作品牌 20 000 多个，独家合作品牌超过 2 200 个，会员超过 3 亿。作为最早进入时尚细分市场的电商平台之一，唯品会除了主打品牌与折扣的卖点，还提供精选导购的价值，并有效融合金融和物流服务，

26 数据来源：唯品会公司。

获得消费者、尤其是时尚女性人群的认可。2017年6月，唯品会进行品牌升级，将沿用多年的定位语“一家专门做特卖的网站”升级为“全球精选 正品特卖”，在特卖基础上着重强调全球精选、正品保证，并在移动端、智能技术、物流环节等方面开展创新升级。从2015年第三季度到2017年第二季度，唯品会连续8个季度稳居行业第三，成为全球最大的特卖电商、中国第三大电商平台[27]。

2. B2B类垂直平台加速向生产领域拓展

垂直电商平台成为企业电商发展的突破口，通过发挥互联网高效连接供需、降低信息不对称的功能，帮助企业化解产能过剩、流通成本过高、有效供给不足等问题。国内钢铁、石油化工、煤炭、有色金属、汽车等行业的企业电商发展迅速，形成了一批百亿元乃至千亿元级的B2B类垂直电商交易平台。构建垂直电商平台和深度参与垂直平台体系，已经成为传统工业企业转型的重要途径之一。“一呼百应”等原材料采购B2B电商，将原材料的供应商与采购商通过平台直接对接，提高了交易效率，降低了下游采购商的成本；钢铁行业电商找钢网，将传统的钢铁交易环节缩短，大幅提升了供应链效率；煤炭供应链企业瑞茂通，通过以煤炭交易为主的大宗商品电子交易平台易煤网，打通上下游完整供应链服务环节。

3.3.3 跨境电商蓬勃兴起

随着对外开放水平的不断提高和国内消费能力的提升，跨境电商成为电子商务发展的重要方向之一。一方面，各级政府大力为跨境电子商务创造贸

27 数据来源：易观智库2015年第3季度至2017年第2季度发布的《中国网上零售B2C市场报告》。

易便利：2015 年 3 月和 2016 年 1 月，国务院先后批准设立杭州、天津等 13 个跨境电子商务综合试验区，为推动全国跨境电子商务健康发展创造更多可复制推广的经验。2016 年，财政部、海关总署、国家税务总局联合发文《关于跨境电子商务零售进口税收政策的通知》和《跨境电子商务零售进口清单》，推动跨境电商行业规范化、机制化、规模化发展。另一方面，大型电商企业也在加快拓展跨境电商业务，2016 年阿里巴巴天猫“双十一”活动吸引了全球 235 个国家和地区参与其中。在政策支持和市场驱动的共同作用下，我国跨境电商进入高速发展阶段，2016 年中国海淘用户规模达 0.41 亿，增速达 78.3%，中国进出口跨境电商（含零售及 B2B）整体交易规模达到 6.3 万亿元，同比增长 23.5%[28]。2016 年，13 个综试区跨境电商进出口超 1 600 亿元人民币，翻了 1 倍多[29]。未来，随着“一带一路”建设的深入推进，跨境电子商务业务将迎来更加广阔的发展空间。

3.3.4 农村与县域电商蒸蒸日上

近年来，国家陆续出台支持农村与县域电商发展的政策，从试点示范、鼓励创新、开拓渠道、金融服务等各方面提供支持，“农产品上行”和“消费品下行”双向流通渠道基本建立，县域网商数量从万级发展到百万级，形成了遂昌模式、沙集模式等具有区域特色的电商模式，电子商务正成为推进城乡协同发展、促进县域经济转型、助力精准扶贫的重要途径。2016 年，我国农村网络零售市场交易规模达到 8 945.4 亿元，占全国网络零售总额的

28 数据来源：商务部、海关总署（艾媒咨询《2016—2017 中国跨境电商市场研究报告》整理）。

29 数据来源：王炳南，十二届全国人大常委会专题讲座第二十九讲 《国内外电子商务的现状与发展》。

17.4%，淘宝村[30]数量达到1 311个，同比增长68%；淘宝镇达到135个，比2015年的71个增长达90%，创造超过84万个就业岗位[31]。

专栏9 电商发展的遂昌模式与沙集模式[32]

遂昌模式：遂昌县地处浙江省丽水市，在政策环境的催化下，通过促进地方农产品加工业的电子商务化发展，逐渐形成了“电子商务综合服务商+网商+传统产业”的县域电商模式。遂昌模式的核心是本地化的电子商务“服务商”，即2010年返乡创业者潘东明、遂昌团委和当地企业共同创建的“遂昌网商协会”及其下属的“网店服务中心”，其主要功能包括整合可售货源、组织当地网商等分销群体以及统一仓储与发货服务，并且建立了农林产品的产销标准、商品产销制作统一数据包，降低了网商的技术门槛和资金门槛，让上游的生产端和下游的销售端专注于擅长的工作。遂昌模式在“九山半水半分田”的典型山区县的成功，为电商基础较弱、小品牌与小网商较多的广大县域提供了典型示范，也吸引了各级领导与大批专家学者的到访与调研，并使遂昌成为中国农村电子商务调研基地。

沙集模式：“沙集模式”起源地在江苏省徐州市沙集镇东风村，其特点是不依赖于当地的传统产业，“无中生有”发展电子商务。2006年开始，以孙寒为首的“电商三剑客”开始在沙集尝试做简易家具电商。简易家具行业的技术门槛、资金需求与资源整合难度相对较低，板材供应商、加工厂、网店、仓储包装等环节在当地产生了社会化分工的模式，并经由大量“农民网商”的网店卖给消费者。在沙集模式中，当地农民网商成为积极对接信息、整合资源的主导者，并通过不

30 淘宝村：指电子商务年交易额达到1千万元以上的村庄。

31 数据来源：阿里研究院，《中国农村电商发展报告》（2017年5月）。

32 数据来源：阿里研究院，《遂昌模式研究》和《“沙集模式”调研报告》。

断裂变式复制，迅速壮大产业集群；以淘宝网为代表的电商平台，成为重要的市场化公共电子商务基础设施；物流、电信、供电、信贷和政府部门则为农村电商产业生态创造了良好的发展环境。沙集模式的成功，对于工农业基础都较为薄弱的中西部地区而言具有明显的示范效应和激励作用。

1. 农村与县域电商拓展了农民增收新途径

传统农业产销面临农产品价格下行和生产成本居高不下的双重压力，电子商务成为优化农产品供应链体系的利器，实现供需直接连接，加快传统农业转型升级。在渠道上，淘宝、天猫、京东、苏宁等大型电商平台介入农产品销售，顺丰优选、本来生活等农产品电商成功搭建了 B2C 电子商务平台，带来品牌化打造、精细化操作等现代农业运作方式。农产品产地也纷纷自建电商平台、微分销平台，与实体店面结合，形成线上线下融合的多元分销体系。在供应端，电商企业开始介入货源的组织与挑选，天猫农业、天猫国际等平台实现基地直采模式，甘肃成县农产品、阳澄湖大闸蟹等区域产品主动对接互联网集采，极大推动货源的商品化与标准化。在消费端，区域公共品牌建设和追溯品控体系的建立，增强了消费者对农产品质量的信任，愿意支付农产品品牌溢价。2016 年，仅阿里巴巴平台的农产品交易额就超过 1 000 亿元，蔬菜成为增长最快的农产品，同比增速高达 151%[33]。从 2012 年到 2016 年，我国农产品网络零售交易总额从 200 亿元增长至 1 589 亿元，增长近 8 倍[34]。

33 阿里研究院：从“客厅革命”到“厨房革命”——阿里农产品电子商务白皮书（2016）。

34 数据来源:国家发改委门户网站，《数字经济引领经济增长新时代》（2017 年 10 月 12 日）。

2. 消费品下行丰富农村消费市场

农村网购的快速发展得益于消费品下行渠道的完善，充分激发了农村市场消费潜力。自2014年以来，商务部会同财政部、国务院扶贫办，安排中央财政资金84亿元，以中西部地区为主，在27个省（自治区、直辖市）的496个县开展“电子商务进农村”综合示范，建设了“县级电商服务中心、乡镇电商服务站、村级电商服务点”三级农村电子商务公共服务网络，有效畅通农村的消费品下乡渠道。京东、苏宁、淘宝、邮政等企业借助农村淘宝、赶街网、农村E邮平台，在1 000多个县建设了40万个电商村级服务点，农村成为电商企业“渠道下沉”的主战场。2016年，农村网购市场规模达4 823亿元，同比增长36.6%，预计2017年将突破6 000亿元[35]。

3.4 互联网金融发展与规范并重

互联网金融的蓬勃发展，为金融服务实体经济注入新动能、提供新方案、实现新跨越。第三方在线支付、尤其是移动支付市场快速崛起，互联网技术不断提升金融服务质量和效率，促进普惠金融、网络借贷、网络众筹等新业态快速发展。面对互联网金融发展中伴随产生的非法集资、虚假承诺、信息泄露等问题，我国政府出台《关于促进互联网金融健康发展的指导意见》等规章制度，积极开展互联网金融风险专项整治，进一步健全互联网金融监管体系，推动互联网金融行业健康有序发展。

35 数据来源：中国电子商务研究中心，《2016 年度中国网络零售市场数据监测报告》。

3.4.1 在线支付服务体系不断完善

支付清算作为传统的金融产业，通过与互联网深度融合，已经成为最具创新活力和发展前景的行业之一，尤其伴随着手机支付业务的不断普及，在线支付已经深刻改变了人们的消费习惯和生活方式。

1. 在线支付市场逐年扩大

为解决电商交易中的供需信任问题，服务于淘宝网的支付宝在 2004 年正式成立运行，我国第三方支付市场开始真正起步。经过 10 多年的发展，在线支付已成为中国金融支付体系的重要组成部分。截至 2016 年，PC 端与移动端合计的第三方在线支付市场规模突破 78.7 万亿元（见图 3-22）。截至 2017 年 6 月，我国使用网上支付的用户规模达到 5.11 亿（见图 3-23），网民使用网上支付的比例从 64.9%提升至 68.0%。

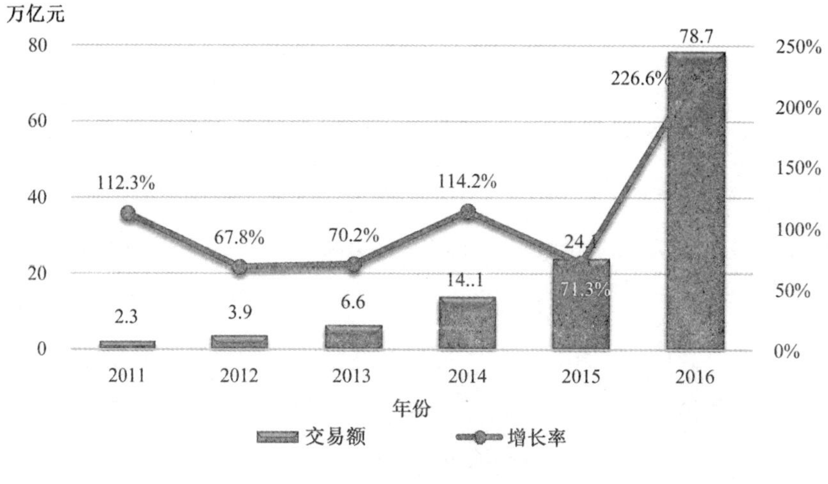

图 3-22　2011—2016 年中国第三方在线支付交易规模及增长率

（数据来源：艾瑞咨询，《2017 年中国第三方支付市场监测报告》）

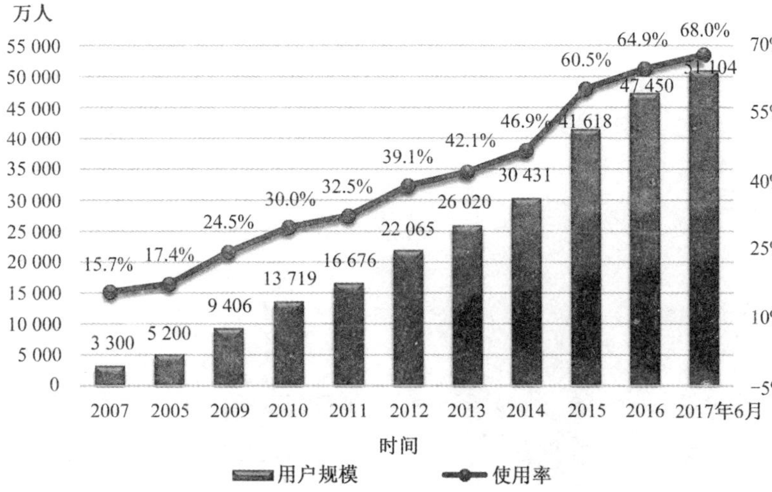

图 3-23 2007—2017 年 6 月中国网上支付用户规模和使用率

（数据来源：CNNIC，第 39、40 次《中国互联网络发展状况统计报告》）

2. 移动支付市场快速扩张

近年来，移动支付市场以惊人的速度扩张，截至 2017 年 6 月，手机支付用户规模已达到 5.02 亿，年增长率为 31.2%，网民手机网上支付使用比例提升至 69.4%[36]。支付宝于 2008 年 2 月推出手机支付业务，并于 2013 年完成手机支付超 27.8 亿笔、金额超过 9 000 亿元，成为全球最大的移动支付公司。2014 年 1 月，微信"抢红包"掀起热潮，仅当年春节期间就吸引超过 800 万用户领取 4 000 多万个红包，极大促进移动支付应用推广。微信财付通与支付宝逐渐形成行业"双巨头"格局，合计占据 90%以上市场份额。2016 年，我国移动支付交易规模已超 81 万亿元，是全球最大的移动支付市场。以腾讯、阿里巴巴为首的中国在线支付企业正在加速全球化布局，目

36 数据来源：CNNIC,第 40 次《中国互联网络发展状况统计报告》。

前，支付宝、微信支付已经覆盖 50 多个海外国家和地区，跨境活跃用户超过 2 亿[37]。2016 年中国第三方移动支付市场份额如图 3-24 所示，2011—2016 年中国第三方在线支付交易规模及增长率如图 3-25 所示。

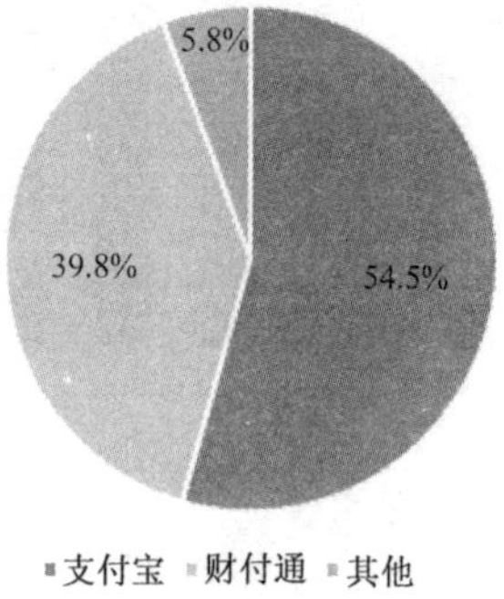

支付企业	交易规模（亿元）
壹钱包	4 207.7
联动优势	2 095.6
连连支付	2 079.0
京东支付	2 068.2
快钱	1 592.4
易宝支付	1 439.0
银联商务	1 242.7
苏宁支付	799.7

图 3-24　2016 年中国第三方移动支付市场份额

（数据来源：灵核产业研究院，2017 年第二季度第三方移动支付市场规模报告）

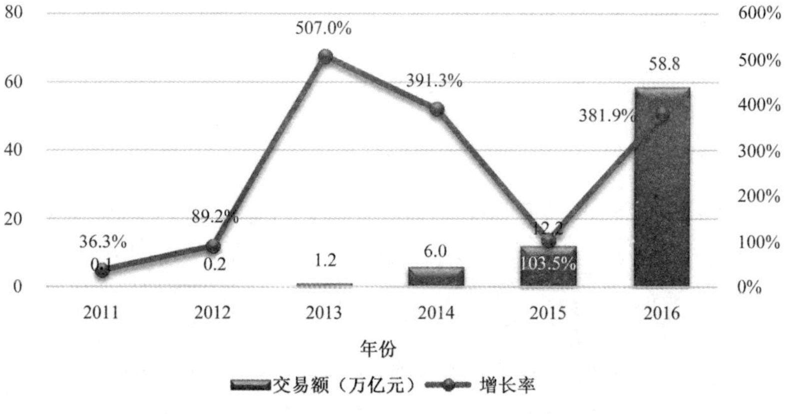

图 3-25　2011—2016 年中国第三方在线支付交易规模及增长率

（数据来源：艾瑞咨询，《2017 年中国第三方支付市场监测报告》）

37 数据来源：根据蚂蚁金服、腾讯公司披露数据整理。

3. 线下支付加快渗透公众生活

经过近年来的市场培育和场景开发，我国已经形成丰富的线下支付场景，二维码、NFC（近场支付）等手机支付技术产生的巨大便利性，对消费者日常支付方式和商户的经营方式带来翻天覆地的变化。移动网上支付在一定程度上已经取代实物钱包，消费者在饭馆、超市、便利店等线下实体店使用移动网络支付工具习惯初步养成，并快速向三四线城市渗透，"无钱包"出行时代悄然开启。网络支付给用户带来购物环节便捷的同时，帮助商家降低收单成本、解决现金管理带来的不便，提升门店消费服务体验。截至2016年底，网民在线下实体店购物时使用手机支付结算的比例已达50.3%，并且线下支付应用拥有较强的下潜力度，四、五线城市分别到达43.5%和38.0%，农村地区使用率已达31.7%[38]。

3.4.2 金融服务网络化快速推广

互联网金融服务具有突破时空限制、打破服务门槛和提升用户体验的优势，彻底改变了传统金融服务模式。1999年9月，招商银行率先在中国全面启动"一网通"网上银行服务，并经中国人民银行批准，成为首家开展网上个人银行业务的机构。在此后的十几年里，行业规模持续扩张、产品服务不断创新、普惠金融应运而生，推动我国互联网金融服务快速崛起。

1. 金融行业利用互联网不断创新服务模式

银行、证券、保险、基金、融资租赁等传统金融服务加快与互联网融合，

38 数据来源：CNNIC，第39次《中国互联网络发展状况统计报告》。

基于云计算、大数据、社交网络的创新性互联网金融产品和服务快速涌现，不断丰富投融资渠道与方式。2014 年，互联网银行开始起步，腾讯旗下微众银行、蚂蚁金服旗下网商银行筹建获批，探索利用大数据技术开展信用评级。2015 年，蚂蚁金服旗下的芝麻信用首次测试个人征信系统，腾讯主推“微粒贷”内部上线，“京东白条”升级到“白条+”，“百度有钱”与清华 MBA 合作切入教育信贷。

2. 普惠金融服务助推小微创新创业

“普惠金融”连续三年出现在政府工作报告中，并被写入“十三五”规划。互联网金融机构在完善信用数据、降低交易成本方面具有天然优势：通过互联网技术集聚海量交易规模，促进单笔交易边际成本趋向于零，大幅降低借贷市场准入门槛；依托大数据挖掘，精准厘定和动态调整授信额度，解决信贷双方信息不对称难题。2016 年以来，蚂蚁金服、百度、京东等互联网企业纷纷与五大国有银行牵手，线上线下渠道全面开展合作，有效覆盖传统金融的部分服务盲区，提升金融服务的覆盖率、可得性、针对性和满意度。惠普金融服务一方面让中小微企业从中受益，另一方面又惠及“金融弱势群体”，帮助农民、城镇低收入人群、贫困人群和残疾人、老年人等及时获取价格合理、便捷安全的金融服务。2016 年，支付宝、腾讯财付通等非银行支付机构在互联网支付业务中交易金额占比仅为 2.5%，但支付笔数占比超过 59%，服务了大量小微客户；截至 2017 年 5 月，阿里巴巴蚂蚁金服累计为 600 多万家小微企业和个体创业者提供了超过 1 万亿元的信用贷款。

3.4.3 网络借贷与众筹发展不断规范

近年，网络借贷和网络众筹行业在快速崛起，同时也带来诸多困扰与风险。为打击行业违法行为、引导企业合规经营、促进行业长期稳定健康发展，相关监管政策陆续出台，网络借贷和众筹逐步走向规范化发展。

1. 网络借贷市场告别野蛮生长

自2013年起，我国网络借贷市场先后经历了起步期和爆发式增长阶段，但到2015年，行业被爆多起跑路事件，网贷平台风险控制与贷后管理亟须加强。随着《网络借贷信息中介机构业务活动管理暂行办法》等规范政策的出台，网络借贷市场进入规范发展轨道。新增平台数量在2014年末达到顶峰后逐渐下降，截至2016年年底，网贷平台总数稳定在5 400家左右（见图3-26），成交额则升至6 526.21亿元，是2014年同期的6.85倍；网贷综合利率水平为9.68%，年均下降3.71个百分点，高利率、高风险的投机型平台和产品热潮已经退去。

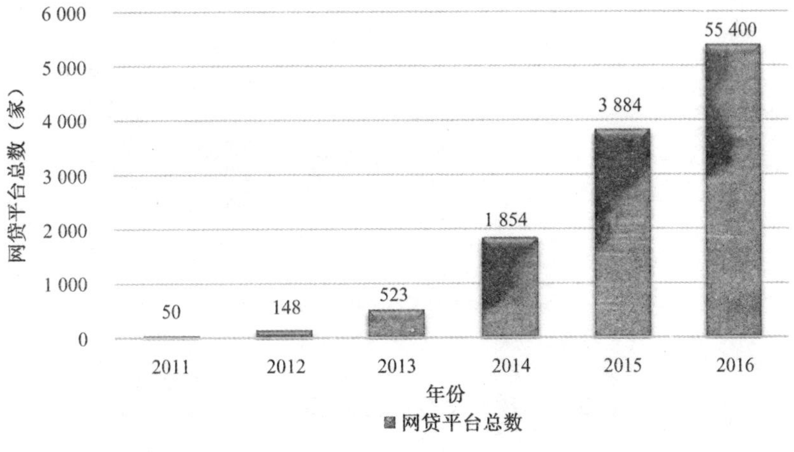

图3-26　2011—2016年我国网贷平台总数及增长率

（数据来源：智研咨询《2017—2022年中国网贷平台市场深度调研及投资前景分析报告》）

2. 众筹行业走向规范化

2011 年 7 月国内首家众筹网站“点名时间”正式上线，成为我国网络众筹市场起点。众筹具备为中小微创业公司筹资、筹智、筹人等先天优势，吸引了大批创业者投身其中。然而，由于涉及领域众多、缺乏法律规范、监管政策不明确、业务运作混乱等问题，行业发展存在较大风险。从 2015 年下半年开始，《互联网金融风险专项整治工作实施方案》和《关于促进互联网金融健康发展的指导意见》等陆续出台，在鼓励和支持行业发展的同时，持续加大众筹行业规范力度。截至 2016 年年底，全国共有众筹平台 752 个（见图 3-27），其中正常运营的 532 个；共发起众筹项目 58 605 个，其中成功项目 48 437 个，实际融资额 217.43 亿元[39]。

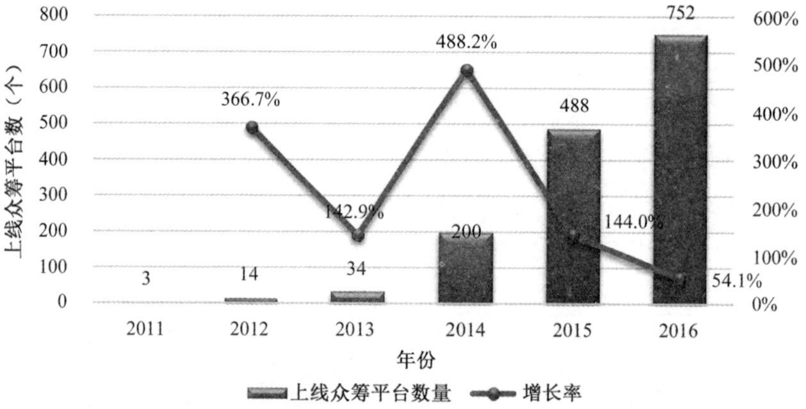

图 3-27 2011—2016 年我国上线众筹平台数量及增长率

（数据来源：人创咨询，《2016 中国众筹行业发展年报》）

39 数据来源：人创咨询，《2016 中国众筹行业发展年报》。

3.5　共享经济成为新的增长点

伴随着信息技术及其创新应用的快速发展，共享经济[40]呈现井喷式发展。2016 年被称为“中国共享经济元年”，共享单车、共享房屋、共享物流、共享金融等模式快速崛起，成为匹配供需、优化资源配置的全新方式。2016 年 6 月《中国互联网分享经济服务自律公约》发布，2016 年 7 月《网络预约出租汽车经营服务管理暂行办法》出台，共享经济发展的政策保障、信用环境和行业自律等各个方面正在逐步走向规范化。

3.5.1　市场规模爆发式增长

当前中国共享经济主要集中在大众消费和服务领域，其中交通出行、房屋住宿、医疗服务等领域表现最为突出（见表 3-1）。在交通出行方面，滴滴、优步等网约车不断发展，截至 2017 年 6 月，我国网约出租车用户规模达到 2.78 亿，增长率为 23.7%[41]；共享单车发展迅速，2017 年共投放单车 2 000 万辆，用户规模达 1.06 亿[42]。在共享住房方面，爱彼迎（Airbnb）、途家、小猪等成为民宿租赁预订服务的领军平台，2016 年住房分享市场交易额约达 243 亿元，比上年增幅高达 131%，共享房源数量达到 190 万套、参与人数

40 共享经济，有时又称为分享经济，英文名称 Sharing Economy，是指将社会海量、分散、闲置资源，平台化、协同化地集聚、复用与供需匹配，从而实现经济与社会价值创新的一种新经济、新业态。

41 数据来源：CNNIC，第 40 次《中国互联网络发展状况统计报告》。

42 数据来源：中国自行车协会数据。

超过 3 500 万人。在医疗共享市场方面，2016 年交易额达 155 亿元，约 256 万人参与提供服务，超过 2 亿消费者参与医疗共享活动。在知识付费方面，知乎、果壳、喜马拉雅 FM、得到、荔枝微课及其他知识付费平台相继出现，知识付费用户达到近 5 000 万人。2016 年中国共享经济的市场交易额约为 34 520 亿元[43]，同比增长 103%。预计到 2020 年，共享经济交易规模占 GDP 比重有望达 10%以上；共享经济在未来十年可能成长为颠覆性的经济力量，预计 2025 年全球共享经济的市场价值将达到 3 350 亿美元[44]。

表 3-1　中国共享经济重点领域市场规模

领域	2015 年交易额（亿元）	2016 年交易额（亿元）	增长率（%）
知识技能	200	610	205
房屋出租	105	243	131
交通出行	1 000	2 038	104
生活服务	3 603	7 233	101
生产能力	2 000	3 380	69
医疗分享	70	155	121
资金分享	10 000	20 863	109
总计	16 978	34 522	103

（数据来源：国家信息中心分享经济研究中心&中国互联网协会分享经济工作委员会，《中国分享经济发展报告（2017）》）

3.5.2　有效促进“双创”发展

共享模式催生交通出行、智能终端应用、家庭生活等各个服务领域的新

43 数据来源：国家信息中心分享经济研究中心、中国互联网协会分享经济工作委员会，《中国分享经济发展报告（2017）》。

44 数据来源：普华永道预测数据。

模式，成为市场创新和跨领域创业的典范，让创新创业活动门槛更低、成本更小、速度更快、参与更广。目前，共享经济领域已出现超过 8 家上市 / 挂牌企业，这些企业从初创到上市/挂牌平均历时 66 个月，领先于互联网全行业的 84 个月[45]；在全球 186 家“独角兽”企业中，中国公司占 42 家，具有典型共享经济属性的公司就有 15 家，占中国独角兽企业总数的 35.7%[46]。共享经济领域“双创”企业的高成长性也得到资本市场密切关注，2016 年，共享经济企业融资规模达 1 710 亿元[47]。

专栏 10 共享单车

共享单车是指企业在校园、地铁站点、公交站点、居民区、商业区、公共服务区等提供自行车单车共享租赁服务，最大化利用公共道路通过率，是一种以分时租赁模式为基础的新型共享经济。2016 年，共享单车概念在国内迅速火爆，当年中国共享单车市场整体用户数量已达到 1 886 万[48]。据不完全统计，尽管经过了几轮优胜劣汰，但目前全国仍有共享单车运营企业超过 20 家。2014 年成立的 ofo、2015 年成立的摩拜单车是其中的典型代表，两者分别和蚂蚁金服及微信钱包合作，迅速抢占国内市场。2017 年，ofo、摩拜等企业更是开始进军新加坡、美国、澳大利亚、日本等海外市场。

与“有桩”的公共自行车相比，随时取用和停车的“无桩”理念给市民带来了极大便利的同时，也导致“乱占道”现象更加普遍，甚至阻挡消防通道、产生安全隐患，加剧城市空间管理困难。为此，2017 年 5

45 数据来源：腾讯研究院，《2017 分享经济报告》，2017 年 5 月 27 日。

46 数据来源：CBInsight 分析数据，《中国分享经济发展报告（2017）》整理。

47 数据来源：国家信息中心分享经济研究中心、中国互联网协会分享经济工作委员会，《中国分享经济发展报告（2017）》。

48 数据来源：中国电子商务研究中心监测数据。

月 7 日，中国自行车协会共享单车专业委员会宣布成立，并将参与制定《共享单车团体标准》。杭州、上海、北京、郑州等城市纷纷出台意见，促进共享单车规范发展。2017 年 8 月 3 日，交通运输部等 10 部门联合发布了《关于鼓励和规范互联网租赁自行车发展的指导意见》，新政提出共享单车平台要提升线上线下服务能力，并明确了规范停车点和推广电子围栏等规范化措施。

3.5.3 创造大量就业机会

共享经济的快速成长创造了大量的灵活就业机会，2016 年我国参与共享经济活动的总人数超过 6 亿人，参与提供服务者人数约为 6 000 万人，比上年增加 1 000 万人[95]。滴滴出行公布的报告显示，2016 年滴滴出行平台为社会创造了 1 750 万个灵活就业机会，其中 238.4 万人来自去产能行业。在共享住房领域，小猪、途家、住百家等平台带动直接和间接就业人数估计超过 200 万人。预计到 2020 年，共享经济提供服务者的人数有望超过 1 亿人，“公司+员工”的社会组织方式将越来越多地被“平台+个人”所替代。

专栏 11　滴滴出行

2012 年 7 月北京小桔科技有限公司成立，同年 9 月滴滴打车上线；2015 年 2 月，快的打车与滴滴打车宣布进行战略合并；2015 年 9 月更名为滴滴出行，2015 年 10 月上海市交通委正式向滴滴快的专车平台颁发网络约租车平台经营资格许可，成为国内第一张专车平台的资质许可。

目前，滴滴出行已涵盖出租车、专车、快车、顺风车、代驾及大巴等多项业务，在中国 400 余座城市为近 3 亿用户提供服务。2015 年全年，滴滴宣布平台共完成 14.3 亿个订单，超过 Uber 成立 6 年累计的

10亿订单数，成为全球最大的一站式出行平台，以及全球仅次于淘宝的第二大在线交易平台。滴滴预期未来以电动化、自动化、共享出行为基础的“交通3.0”将重新定义交通出行，超级拼车、智能红绿灯和智能驾驶等模式将彻底解决拥堵、交通事故、尾气污染和停车难等问题[49]。因此，滴滴将致力于以共享经济实践响应中国互联网创新战略，与不同社群及行业伙伴协作互补，运用大数据驱动的深度学习技术，解决中国的出行和环保挑战。

3.6 “互联网+”改造提升传统行业

2015 年 7 月，国务院印发《关于积极推进“互联网+”行动的指导意见》，促进互联网与第一、二、三产业深度融合发展，推动中国产业迈向中高端。

3.6.1 “互联网+服务”激活双创生命力

1. 互联网助推生活性服务业稳步增长

在“互联网+交通”方面，互联网催生了综合交通运输服务新体系，ETC、一卡通、电子订票等服务模式极大便利日常出行，全国联网的 ETC 系统服务近 4 000 万用户[50]，导航电子地图实时提供基于位置服务的各类智能出行信息。在“互联网+旅游”方面，中国在线旅行预订市场增长迅猛，截至 2016

49 人民网-IT 频道，滴滴总裁柳青：未来出行将“告别”拥堵和事故，2017 年 6 月 27 日，http://it.people.com.cn/n1/2017/0627/c1009-29366714.html

50 数据来源：交通运输部 2016 年 9 月发布数据。

年末用户规模已近 3 亿、使用率超过 40%（见图 3-28）。在“互联网+教育”方面，2016 年在线教育市场规模达到 1 560.2 亿元，同比增长 27.3%，预计未来几年将继续保持 20%的增速[51]。此外，新技术不断驱动线上线下（O2O）互动融合，基于用户数据的深度挖掘进一步增强服务与需求的匹配，在租车约车、餐饮外卖、家政服务、美容保健、教育培训、车辆维保等众多服务领域，O2O 服务模式日益丰富。

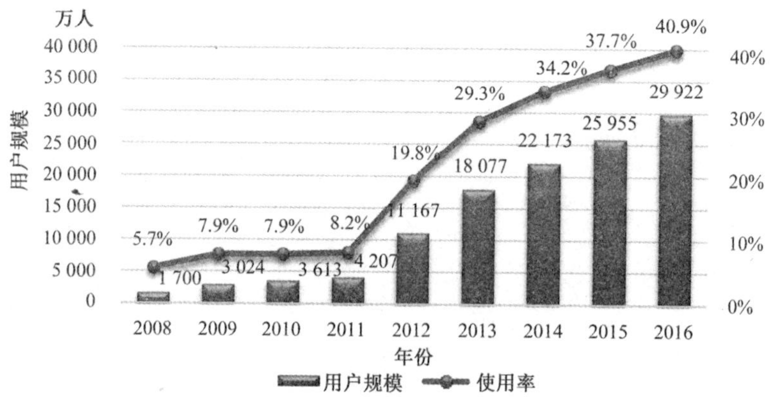

图 3-28　2008—2016 年中国旅行预订用户规模和使用率

（数据来源：CNNIC，《中国互联网络发展状况统计报告》）

2. 互联网推动物流服务业转型升级

网络零售的崛起推动快递业务量的爆发式增长，2010—2016 年，中国快递服务企业业务量从 23.4 亿件增长至 312.8 亿件，业务收入从 574.6 亿元增至 3 974.4 亿元，其中超过 60%的快递业务来自网络购物[52]。通过互联网

51 数据来源：艾瑞咨询，《2017 年中国在线教育平台研究报告》。

52 数据来源：国家邮政局，《2016 年中国快递发展指数报告》。

手段实现的大规模集单，推动物流行业集约化和直线化发展，从而从整体上降低整个社会的物流和仓储成本，解决我国物流行业成本过高的瓶颈。同时，互联网倒逼物流行业转型升级，物流龙头企业纷纷加强物流网络建设与电商合作，提供“异地次日达”、“同城当日达”、“一日多送”等高质量服务，通过实时包裹查询跟踪服务、完善保险制度和追责赔偿条款等措施，不断提升物流服务水平。

3.6.2 “互联网+制造”提升柔性生产力

2015 年 5 月，国务院印发《中国制造 2025》，部署全面推进实施制造强国战略；2016 年 5 月，国务院印发《关于深化制造业与互联网融合发展的指导意见》，提出培育发展制造业与互联网融合的新产品、新模式、新业态为重点，以“互联网+”推动制造业提质增效与转型升级。互联网技术在生产制造环节应用不断深化，显著提升企业在协同研发、精益管理、风险管控、供应链协同、市场快速响应等方面的竞争力。2016 年，我国数字化研发工具普及率、工业企业数字化生产设备联网率分别达到 61.8%和 38.2%[53]；通过“两化融合”管理体系贯标的企业，新产品研发周期缩短 32.6%，产品设计变更率下降 25.7%，生产计划完成率提升 3.7%，用户年投诉次数下降 48.6%，综合效益明显提升[54]。“互联网+”推动柔性制造、个性化定制和服务型制造快速兴起，红领集团、海尔集团、维尚家具、小米科技等不同领域企业纷纷打造顾客直连工厂（C2M）、制造商对消费者（M2C、F2C）等新型网络直销模式，陕鼓、徐工、东方电气等企业加速发展基于互联网的全生命周期管理、远程运维和融资租赁业务，实现从局部流程优化向全业务流程

53 数据来源：工信部部长苗圩十二届全国人大常委会第七次会议上的工作汇报。

54 数据来源：工信部官方网站工作动态，两化融合管理体系贯标工作取得显著成效，2017 年 3 月 7 日。

再造转变、从单项业务应用向多业务综合集成转变。以云设计、云生产、云管理、云试验等为核心的云制造体系加速形成，航天科工航天云网、北京“云链”工业 SaaS 云平台、“徐工工业云”平台等相继上线，佛山市家居行业云制造平台已吸引 3 万多家企业集聚[55]。工业互联网行业的快速发展为实现智能制造提供关键支撑，2016 年中国工业互联网市场规模达 1 896 亿元，同比增长 27.33%，预计在“十三五”期间将保持 25%左右的复合增长率[56]，未来 20 年，中国工业互联网的发展将带动 3 万亿美元的 GDP 增量[57]。

3.6.3 “互联网+农业”助推焕发新活力

近年来，“互联网+农业”相关的国家新政密集出台，推动互联网融合渗透到农业研发生产、市场流通、金融配套等各个环节，促进农产品消费结构升级和附加值提升，助推传统农业焕发新的生命力。在研发生产方面，互联网助推农业技术咨询的普及，并通过新一代信息技术与传统农业管理技术的结合，实现专业规模集约化生产，提高农业生产信息化、智能化水平。目前，农业部已在 8 个省份开展农业物联网试验示范，推介了 426 项技术、产品和应用模式，建成国家农业数据中心、国家农业科技数据分中心及 32 个省级农业数据中心[58]。在市场流通方面，除互联网企业纷纷携技术、资金进军农村电商市场外，行情咨询、市场营销设计、精准物流服务等农业社会化服务不断提升网络化水平。国内最大农村互联网平台“村村乐”致力于农村互联网创业、农业整合营销、农资流通等服务，已经为全国 60 多万个村庄每村

55 数据来源：佛山制造业公共云服务平台“中国在线制造”官网数据。

56 数据来源：赛迪顾问研究成果。

57 数据来源：工信部部长苗圩在全国两会上解读“中国制造 2025”，2015 年 3 月 7 日。

58 数据来源：农业部官方网站，“互联网+”现代农业掀起农业信息化新浪潮，2017 年 1 月 18 日。

建立一个网上门户，注册会员超过1200万人[59]。在农业金融领域，互联网平台在农业产业链各环节提供征信、支付、理财、贷款、保险、融资租赁等服务。截至2017年8月，仅农业金融龙头“农金圈”理财平台的累计交易额就超过106亿元，为3 000余家农资经销商、1万余个规模种植户提供互联网借贷服务[60]。此外，“互联网+农产品质量安全”平台建设推动农产品全产业链环节安全可追溯，“互联网+休闲农业”推动农业观光旅游和农产品体验式消费发展，不断扩大现代农业盈利渠道、农产品利润空间和农业终端市场消费群体规模。

59 数据来源：村村乐官网（http://www.cuncunle.com/company_profile）。

60 数据来源：农金圈发布数据。

第 4 章　电子政务打造透明高效服务型政府

习近平总书记指出，信息是国家治理的重要依据，要以信息化推进国家治理体系和治理能力现代化。中国高度重视电子政务建设发展，自 20 世纪 80 年代以来，先后经历了三个发展阶段：第一阶段是起步发展，以党政机关办公自动化（OA）工程为重点，先后启动“金字工程”“政府上网工程”等多项政务信息化工程。第二阶段是全面启动，党中央和国务院出台一系列政策文件，提出将电子政务建设作为今后一个时期我国信息化工作的重点，鼓励政府先行，带动国民经济和社会发展信息化，电子政务建设工作全面铺开。第三阶段是创新转型发展，中国电子政务积极推进建设目标、建设方式、建设模式的转型发展，创新社会治理方式，打通信息壁垒，促进民生改善，更好利用信息化手段感知社会态势、畅通沟通渠道、辅助科学决策。

4.1　电子政务支撑保障体系不断健全

4.1.1　电子政务网络架构基本建成

1.“金字工程”奠定电子政务网络框架根基

1993 年年底，国家正式启动了“三金工程”，即金桥、金关和金卡工程，极大促进了国民经济的迅速发展。2002 年 8 月，中共中央办公厅和国务院办公厅发布《国家信息化领导小组关于我国电子政务建设指导意见》，规划安排了“两网、四库、十二金”，覆盖了电子政务建设各领域，初步构成我国电子政务建设的基本框架。中国政府以“金字工程”为重要抓手，加快信息化基础设施建设，各领域网络化管理水平得到普遍提升。经过多年发展，金关、金税、金盾等一批国家重点信息应用系统达到世界先进水平，为民服务效率显著提升。截至2016年年底，人力资源和社会保障部社会保障“一卡通”全国持卡人数达到9.72亿，覆盖全国70%人口，31个省、自治区、直辖市和新疆生产建设兵团发行全国统一社保卡，地市覆盖率达96.6%，全国平均开通80项服务目录[1]，各地逐步推开医疗费用即时结算、跨地区持卡应用、养老金领取、社保个人自主缴费等服务，获得社会公众广泛欢迎。

专栏12 金字工程

三金工程，即“金桥工程”、“金卡工程”、“金关工程”。“金桥工程”首先建立国家共用经济信息网。具体目标是建立一个覆盖全国并与国务院各部委专用网连接的国家共用经济信息网。“金关工程”是对国家外贸企业的信息系统实联网，推广电子数据交换技术（EDI），实行无纸贸易的外贸信息管理工程。“金卡工程”则是以推广使用“信息卡”和“现金卡”为目标的货币电子化工程。

“两网、四库、十二金”，即建设和整合统一的由政务内网和政务外网构成的电子政务网络；规划和开发人口基础信息库、法人单位基础信息库、自然资源和空间地理基础信息库、宏观经济数据库的重要政务信息资源；建设和完善政府办公业务资源系统、金关、金税、金融监管（含金卡）、宏观经济管理、金财、金盾、金审、社会保障、金农、金质、金水等重点业务系统。

1 数据来源：人社部网站发布的《2016年全国社会保障卡工作情况大数据报告》。

2. 电子政务网络基础支撑能力明显增强

2007 年 9 月，国家电子政务网络中央级传输骨干网络正式开通，支持政务内网、政务外网建设和运行的基本载体初步建立，标志着统一的国家电子政务网络框架基本形成，为各部门各地区开展业务应用提供了一个安全可靠、资源丰富、管理规范、服务专业的公共平台。经过多年发展，电子政务内网平台和顶层互联建设工作稳步推进，网络支撑保障作用逐步显现，截至 2016 年年底，已建成政务内网平台省级 32 个，副省级 15 个，市县级以下按照“确保安全、按需建设”原则开展建设；其中，浙江、广东和贵州 3 个地方省级网上政务服务平台已经实现省、市、县、乡四级覆盖，8 个地区实现省、市、县三级覆盖[2]。同时，政务外网接入贯通率也不断提高，截至 2016 年 6 月底，各级政务外网共接入政务部门约 14.7 万家，接入终端超过 184.4 万台。其中，有 130 个中央政务部门和单位接入中央级外网平台，省级平台接入率达到 100%，市级平台接入率为 94.6%，县级和乡镇以下接入率分别为 89.5%和 34.1%[3]。

4.1.2 电子政务治理体系和制度规范日益完善

中央网络安全与信息化领导小组成立以来，加大电子政务发展统筹协调力度，从初期的“一把手工程”转向法制化、制度化、专业化的电子政务治理体系。在地方层面，各省份相继成立了网络安全和信息化领导小组，陕西、甘肃等地率先确定由当地网信领导小组作为电子政务统筹部门，有力促进地方电子政务发展。近年来，国家陆续出台政策文件，电子政务建设目标、建

---

2 数据来源：《中国电子政务发展报告（2017）：“互联网+”创新政务服务新模式》。

3 国家信息中心信息化研究部：《信息化研究》2017 年第 8 期《我国电子政务发展现状与“十三五”展望》，http://www.sic.gov.cn/News/258/8150.htm

设方式和系统模式不断明晰。2016 年，“互联网+政务服务”首次写入《政府工作报告》，对政府职能转变和经济社会发展产生深远影响，开拓了中国电子政务发展新模式。《“十三五”国家信息化规划》将“互联网+电子政务”列入 12 项优先行动。2016 年 9 月，国务院印发了《关于加快推进“互联网+政务服务”工作的指导意见》，提高网上服务事项全面性、信息共享程度、可办理率、企业和群众办事便捷程度。为推动电子政务实践有序推进，围绕统筹规划、信息共享、资源优化整合、管理运营机制等内容，制定发布了《关于进一步加强政务部门信息共享建设管理的指导意见》《政务信息资源共享管理暂行办法》《政务信息系统整合共享实施方案》《关于全面推进政务公开工作的意见》等文件，解决长期困扰我国政务信息化建设的“各自为政、条块分割、烟囱林立、信息孤岛”问题。截至 2016 年年底，中办、国办、国家发展改革委等 30 多个部委出台近 100 项电子政务相关规划、管理办法，其中约 43%与业务系统相关，26%与安全保障相关，13.9%与门户网站相关，12%与标准规范相关[4]，为电子政务健康协调可持续发展创造了良好的环境。

4.2 电子政务服务日益丰富完善

4.2.1 政府网站的公共服务作用愈发突出

1999 年，中国“政府上网工程”正式启动，各地区各部门积极推进政

4 数据来源：《2017 年中国电子政务发展报告》。

府网站建设，向社会公众提供在线服务。目前，中国 96%的政务网站提供网上在线办事服务，基本实现个人和企业用户需求全覆盖，政务事项网上全程办事率达到 73.3%[5]。政务网站围绕社会公众实际办事需求，整合各部门信息和服务资源，提供跨部门、跨领域的全流程在线服务，内容涵盖生活、医疗、社保、住房、交通等全生命周期服务需求。2016 年，中国政府印发《关于加快推进“互联网+政务服务”工作的指导意见》，提出到 2017 年年底前，建成一体化网上政务服务平台，全面公开政务服务事项。2020 年年底前，建成覆盖全国的整体联动、部门协同、省级统筹、一网办理的“互联网+政务服务”体系，大幅提升政务服务智慧化水平。面向个人的在线服务包括生育、户籍、教育、文化、就业、兵役、婚姻、纳税、社保、交通、医疗、邮政、出入境、法律援助等方面；面向企业的在线服务覆盖登记设立、检查审查、投资纳税、劳动保障等与企业紧密相关的各类服务。

专栏 13　中国政府网设立“我向总理说句话”栏目

2014 年，中国政府网开始设立“我向总理说句话”栏目，开展网民建言征集活动，成为政府提供政务服务新平台，受到广大网民欢迎和好评。据统计，从 2014—2016 年，留言总量从 96 774 条跃升至 404 090 条，同比增长约 96.5%。对于公众的每一条留言，政府工作人员都认真对待，并将意见和建议进行汇总整理，专门研究，一些好的建议直接被送到了总理的办公桌上。在 2016 年的 40 多万条留言中，有 2071 条留言提交给国务院领导和政府工作报告起草小组。其中，101 条留言内容与政府工作报告高度契合，栏目已成为网民与政府沟通的重要桥梁。例如，2016 年 7 月 15 日，来自北京的 40 岁网民洪岩留言：别让出差人员“背着营业执照”开票，建议简化营改增办理手续。“敏儿 8068”等 1 974 位网民也就“营改增全面推行”留下了建议。2016 年

5 数据来源：CNNIC。

12月29日，时任财政部部长肖捷明确表示，截至11月底，全面推开营改增试点累计减税3 387亿元。12月27日，税务总局局长王军表示，2017年营改增的减税规模会进一步扩大。

专栏14 广州市的“一窗式”和佛山市的“一门式”服务改革模式

2016年3月，广东省人民政府办公厅印发的《关于在全省推广一门式一网式政府服务模式改革的实施方案》提出，全面梳理和公开公共服务事项目录，坚决砍掉各类烦琐的证明和手续、大力推进办事流程简化优化和服务方式创新；制定行政许可和服务事项的统一申办受理标准，省、市两级要以一网式服务为主、拓展建设网上办事大厅统一申办受理平台，与部门审批系统无缝对接，建立前台综合受理、后台分类审批、统一窗口出件的服务模式，探索法人行政许可和服务事项一门一网办理模式，推广自然人行政许可和服务事项“一门在基层”通用模式。

4.2.2 政务新媒体推动社情民意沟通渠道便捷多样

1. 政务新媒体数量实现大幅度增长

中国政府积极顺应移动互联网与新媒体发展形势，将微博客、微信、移动客户端等作为联系群众、服务群众的重要渠道，积极推进政务新媒体快速发展。2011年，国家网信办召开积极运用微博客服务社会座谈会。2015年2月，国家网信办召开政务新媒体建设发展经验交流会，要求大力推动各地党政机关、企事业单位和人民团体运用即时通信工具开展政务信息工作，“两微一端”政务新媒体的概念在会上被首次提出，政务新媒体迎来快速发展。截至2017年6月，经过新浪平台认证的政务微博账号达到17万个，政务类微信公众账号达 51.4 万个，各级党政机关开通政务头条号账号达到

34 083 个[6]，政务微博、政务微信、政务 APP、政务头条号在发布权威信息、开展政民互动、引导网络舆论方面发挥了重要作用。人民网分析显示，在 2016 年 1—11 月的 600 多起舆情案例中，政府回应率达到 87%，对 57%以上的事件政府在事发 24 小时之内首次响应，对 73%的事件政府部门在 48 小时之内首次回应，41%的事件通过政务新媒体作出回应[7]。政务微博总体分布如图 4-1 所示。

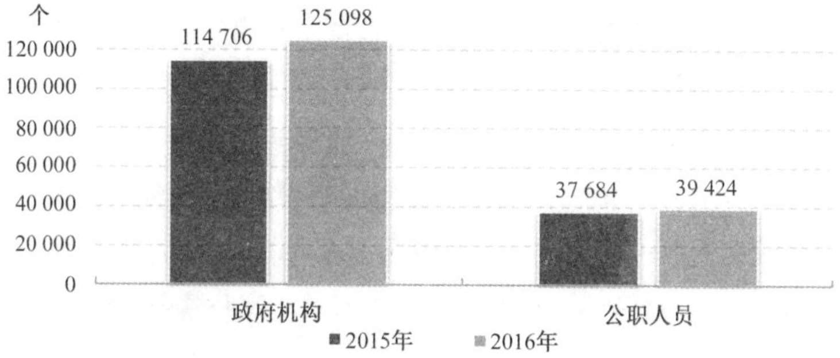

图 4-1　政务微博总体分布

（数据来源：新浪微博）

2．政务新媒体开启掌上政务新渠道

政务新媒体的发展加快了政务信息流动速度，使政务服务内容日趋丰富多元，风格更加亲民，互动更加明显，活跃度持续提升。政府部门与互联网企业加强政务新媒体建设合作，推进政务新媒体普及化、便民化。截至 2016 年年底，全国已有 362 个城市借助微信平台开通了在线城市服务，覆盖用户超过 4 亿，基于微信平台可以提供社保、交通、税务、教育、医疗、水电煤

6 数据来源：CNNIC《第 39 次中国互联网络发展状况统计报告》。
7 《2016 年全国政务舆情回应指数评估报告》，人民网新媒体智库。

气缴费等领域 50 多类在线服务[8]。阿里支付宝支持全国 27 个省份的民生服务[9]，通过运行在阿里云计算平台上的“政务超市”，市民可以像逛淘宝一样“办政事”，通过该平台浙江省市县三级政府 6 万余个审批事项可一网办理，并且实现在线支付缴费[10]。

4.3　信息资源共享推动政务服务现代化

中国政府高度重视并积极推进信息资源开放共享，将政务数据的开发利用作为释放数字红利、促进经济发展的重要手段，积极探索政务数据资源开发利用新实践、开放共享新模式，实现建设开放政府、透明政府，促进公平正义、推动包容发展的电子政务建设发展目标。

4.3.1　信息公开与数据开放推动建设透明高效政府

1. 加快政务信息公开

2008 年 5 月，《中华人民共和国政府信息公开条例》正式实施，中央部委及直属机构、省级政府等公布了依申请公开的信息公开流程，政府信息公开更加规范完善。中国政府网（www.gov.cn）整合政府信息公开目录系统，全面发布中央人民政府制定的各类文件，发布了中央政府各部门的权力清单

8 数据来源：腾讯《“互联网+人社”发展白皮书》。

9 数据来源：阿里巴巴。

10 数据来源：环球网、大河网等对“浙江政务服务网”的综合报道。

（见图 4-2），每个行政许可事项都能够提供设定依据、服务指南、审批流程等内容，对政府各部门依法、透明、规范行政起到重要的督促作用。截至 2016 年 12 月，全国 31 个省（自治区、直辖市）的行政许可事项中，100%的事项已经实现政务服务事项信息的全面公开；80.41%实现了提供相关申请材料的文本可在网上下载的单项服务；62.33%的事项实现在线提交相关申请材料进行预审的双向处理；56.47%的事项在交互服务的基础上，实现了完全在线办理[11]。中国政府通过互联网发布大量社会公众关心的政府公报、会议活动、人事任免、财政预决算、反腐倡廉等关键性信息，在线直播报道领导会议或活动，直播政府及部门会议、新闻发布会，组织直播政府及有关部门重要活动，信息公开取得实质进展。

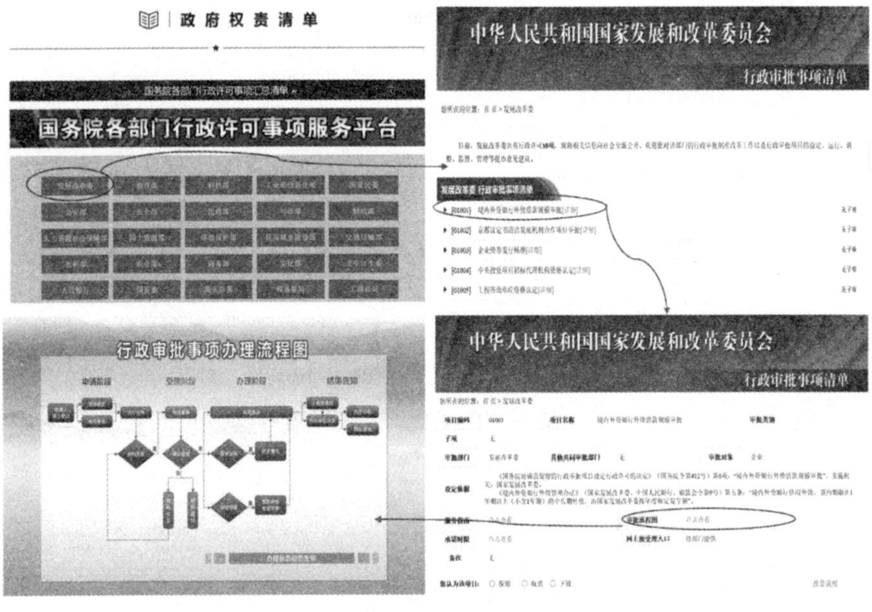

图 4-2　中国政府网发布的部门权力清单

11 数据来源：《中国电子政务发展报告（2017）》。

2. 推动政府部门间信息共享

中国致力于打通政府部门间数据壁垒，通过搭建统一的数据开放平台，满足社会公众获取信息、处理信息、使用信息的需求。2016 年 9 月，国务院印发《政务信息资源共享管理暂行办法》，要求加快推动政务信息系统互联和公共数据共享，充分发挥政务信息资源共享在深化改革、转变职能、创新管理中的重要作用。2017 年《政府工作报告》明确指出，要加快国务院和地方政府信息系统互联互通，形成统一政府服务平台。2017 年 5 月，国务院印发《政务信息系统整合共享实施方案》，提出“政务数据共享和开放在重点领域取得突破性进展”的目标。各地区各部门积极落实信息共享规划要求，加快推进信息共享实践，取得显著成效。截至 2016 年年底，国家人口基础信息库人口基础信息达到 13.5 亿条，人口照片采集饱和度达到 96.11%，实际居住地信息饱和度达到 78.53%[12]。人口信息库基本实现了公安、教育、民政、人社等部门的信息共享，在公共安全、社会保障、宏观调控等领域发挥了巨大作用。北京市小汽车摇号系统，共享了公安、社保、税务、民政等部门信息，避免了重复采集信息，降低了信息采集成本，提高了信息准确率，确保小汽车摇号公平公正公开。

3. 助推“简政放权”部署

一站式政务平台是全面落实“简政放权”要求的重要实践，各级政府深化数据资源开放应用，加大数据整合力度，不断加强资源增值服务，为用户更好地享受政府服务。浙江省“放管服”改革试点，打造统一的人口、法人、电子证照库，实现工商、税务、人社等部门信息共享，截至 2016 年 4 月，共享人口数据达 5 896 万条、法人数据 249 万条，集纳 37 个部门的 860 万

12 数据来源：《中国电子政务年鉴（2016）》。

条电子证照，通过信息共享办理“五证合一”营业执照 56 万多张，建设厅资质审批过程中通过电子证照共享参保信息，让办事企业和群众免于多部门查验，累计达 11 万多次[13]。

4.3.2 新兴信息技术提升政务服务效率

1. 基于云计算的电子政务集约化建设快速普及

电子政务业务应用正从以往的分散建设，向基于云计算的集约化建设模式转变。据统计，截至 2016 年年底，全国已有 15 个省级地方通过“云”实现了机房、存储设备、OA 系统等软硬件资源的集中建设。其中，陕西将 95 个省级部门、11 个地市、107 个县的 1 233 个业务应用系统和数百个数据库部署在政务云平台上，通过集约化建设，节约了 55%的建设资金、60%的运维服务费用。

2. 利用大数据技术推动科学决策

大数据可直观地揭示传统技术手段所不能展现的事务关联，让宏观调控决策更科学，让微观管理服务更精准。中国证监会运用大数据分析对证券交易中的“老鼠仓”、关联交易等行为进行打击，提高了金融、证券监管部门对市场风险、违规交易的跟踪和监管能力，实现监管水平的大幅提升。国家税务总局“金税三期”工程实现全国应用系统大集中，采用统一的数据标准和口径，打通涉税数据跨省市、跨部门流通的关键环节，通过深化大数据共享与利用，大幅度提高税源监管质量水平，并且实现全国办税业务流程的规

13 数据来源：陈立三，《“互联网+政务服务”——浙江实践》。

范，以及各地税务机关执法标准的统一。上海、山西、成都等建立交通综合信息平台，对涉及多个部门、单位的交通数据进行整合，实时监控道路出现的异常拥堵现象，为交警及时处置和公众合理选择出行线路提供依据，为城市规划等提供大数据支撑。

3. 利用人工智能技术提升政务服务智能化水平

人工智能技术在电子政务领域的应用，将使服务从人对人的服务转变为机器对人服务，从面对面的服务转变为不见面的服务，逐步改变传统的政务服务模式。通过自主学习并应用人、地、物等情景信息，提供自动识别和智能预测，目前已在防恐反恐、环境监测、气候预测等领域得到广泛应用。随着智能化、个性化服务需求的激增，人工智能在电子政务领域的应用将更加普及。

第 5 章　网络空间天朗气清

习近平总书记指出，网络空间是亿万民众共同的精神家园，网络空间天朗气清、生态良好，符合人民利益。中国本着对社会负责、对人民负责的态度，大力开展网络内容建设，持续做强网上正面宣传，积极培育健康向上的网络文化，着力推进媒体融合发展，加快推进互联网法治建设，依法加强网络空间治理，使网络正能量更加充沛、网络文化更加繁荣、网络秩序更加规范、网络空间更加清朗，为广大网民特别是青少年营造天朗气清的网络空间。

5.1　网络舆论生态积极健康向上

5.1.1　网络媒体成为重要信息传播渠道

1. 网络新闻信息传播主体不断壮大

网络新闻信息服务经过门户网站、社交媒体和移动媒体三个阶段的发展，传播主体不断发展壮大，成为传播新闻信息的主阵地。1997 年元旦，《人民日报》主办的人民网开通，成为全国第一家中央重点新闻宣传网站。随后，新华社、中央电视台等中央主要新闻单位陆续开通网站，从事网络新闻信息服务。一批地方新闻网站快速发展，网易、搜狐、新浪等商业门户网站相继成立并开通新闻频道。随着移动互联网市场的爆发式成长，以今日头条、澎湃等新闻客户端为代表的移动媒体，以及以新浪微博、微信等为代表的社交媒体发展迅猛，传播影响力不断增强，成为新闻信息传播的重要渠道。目前，中国已经形成以重点新闻网站为骨干，主要商业网站、移动端网络媒体积极参与的媒体格局。截至 2017 年 6 月，具有互联网新闻信息服务许可资质的机构共 297 家，网络新闻用户规模达 6.25 亿[1]（见图 5-1）。

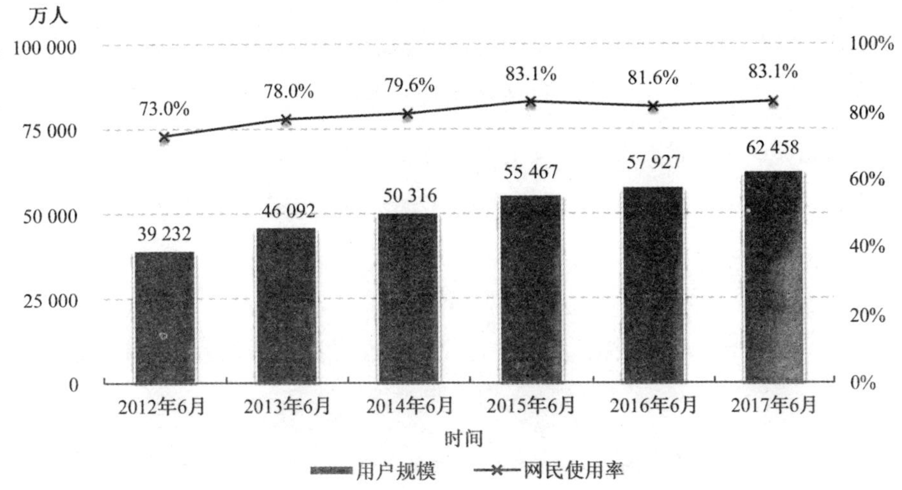

图 5-1　中国网络新闻用户规模和网民使用率

（数据来源：CNNIC，《中国互联网络发展状况统计报告》，2017 年 6 月）

2. 网络媒体传播能力不断增强

互联网新闻用户规模的迅猛增长与技术服务水平的不断提升极大增强

1 数据来源：CNNIC。

了网络媒体的传播能力。网络媒体围绕党和国家事业大局，持续加强网上正面宣传，推出“深改组 1000 天”等一系列重大主题宣传，精心开展网上形势宣传、成就宣传、典型宣传，互联网日益成为宣传学习习近平总书记系列重要讲话精神和党中央治国理政新理念新思想新战略的主渠道。在中国共产党第十九次全国代表大会期间，15 家中央主要新闻网站成为十九大报道的“国家队”，在新闻报道中承担了主力军和领头羊的角色。数据显示，“十九大”相关报道阅读量超过 170 亿，相关视频播放量超过 32 亿[2]；此外，十九大期间创新新闻传播方式，第一次开通微信公众号“十九大新闻中心”，更好地满足了中外记者的采访需求和公众通过移动设备了解党的新闻信息的需求。同时，网络媒体发挥网络传播优势，遵循网上传播规律，在社会民生热点问题、重大自然灾害和突发公共事件报道中发挥重要作用，第一时间发布权威信息，第一时间解读政策措施，及时回应社会关切，解疑释惑、澄清谣言、疏导情绪。

专栏 15 纪念中国人民抗日战争暨世界反法西斯战争胜利 70 周年大阅兵活动报道情况

2015 年 9 月 3 日，北京举行纪念中国人民抗日战争暨世界反法西斯战争胜利 70 周年大阅兵活动，全球 10 亿人同时在线浏览、收看纪念大会和阅兵式，网上相关报道超过 2 000 万篇，点击量超过 8 亿次，跟帖评论约 1 亿条，新浪微博阅兵话题总阅读量达 51 亿次，讨论量 856 万条，微信消息讨论量超过 18 亿条，阅兵视频点击量超 3 亿次[3]。

---

2 数据来源，新浪微博。

3 数据来源：中国青年网，全球 10 亿网民“指尖大阅兵”爱国热潮涌动网络空间，中国青年网北京 2015 年 9 月 4 日电。

3. 网络媒体积极传播中国声音

网络媒体已经成为讲述中国故事、传播中国声音的重要窗口和渠道。网络媒体积极创新对外话语体系，顺应互联网跨国界传播趋势，探索跨文化传播技巧，通过加强议题设置、引入境外编辑人员等方式，运用海外受众乐于接受的方式、易于理解的语言，着力打造融通中外的新概念新范畴新表述，提升传播内容的感染力、针对性和有效性。同时，积极拓展传播阵地，人民网、新华网、央视网、中国网、中国日报网、国际在线等网站实现了以多种语言文字向全球发布新闻信息和视听节目，并积极推出多语言版本的客户端服务。《人民日报》社、新华社、中央电视台、《中国日报》社等机构积极在脸谱（Facebook）、推特（Twitter）、优兔（YouTube）、连我（LINE）、VKontakte（俄罗斯）等境外社交平台上开通账号，账号粉丝数量及关注量累计超过 1.1 亿，网络国际传播覆盖 200 多个国家和地区、10 亿多海外用户，网络内容落地率进一步提高。

5.1.2　网络媒体演进创新推动行业繁荣

1. 媒体融合发展取得重要进展

中国持续推动传统媒体和新兴媒体在内容、渠道、平台、经营、管理等方面的深度融合，打造了一批覆盖广泛、形态多样、手段先进、具有较强影响力的新型传播平台。2014 年，中央全面深化改革领导小组通过《关于推动传统媒体和新兴媒体融合发展的指导意见》，为推动传统出版和新兴出版融合发展提出方向、目标和路径。2016 年 7 月，国家新闻出版广电总局印发《关于进一步加快广播电视媒体与新兴媒体融合发展的意见》的通知，将

“建成多个形态多样、手段先进、具有竞争力的新型主流媒体，打造出数家拥有较强实力的新型媒体集团”作为媒体融合发展目标。此后，人民日报、新华社、中央电视台等主流传统媒体纷纷加强了对于新媒体传播的重视程度，积极布局“两微一端”，打造“中央厨房”模式，成为带动媒体融合发展的主力军。截至 2017 年年初，《人民日报》客户端用户自主下载量达到 1.54 亿，《人民日报》法人微博在人民网、新浪网、腾讯网三大平台上的粉丝总数突破 8 900 万，微信公众号订阅量超过 840 万；新华社客户端累计下载量达到 1.8 亿，各微博账号粉丝总量 4 467 万，微信公众号订阅量 61 万；中央电视台“央视影音”客户端下载量 5.5 亿，“央视新闻”客户端下载量 4 195 万、微博粉丝量 4 700 万、微信公众号订阅量 680 万。

2. 移动新媒体影响力持续增强

近年来，手机报、手机 WAP 网站、移动客户端、官方微博、微信公众账号等移动新媒体不断推陈出新，影响范围越来越大。截至 2017 年 6 月，中国手机网络新闻用户规模达到 5.96 亿人，占比达整体网络新闻用户的 95.4%（见图 5-2），移动端超过 PC 端成为网民获取新闻信息的第一来源，微信朋友圈、微博的使用率也分别达到 84.3%和 38.7%[4]。此外，传统媒体借助移动平台打造手机报新媒体，加快推动传统媒体与新媒体融合进程。截至 2016 年 7 月，全国手机报用户总数超过 3 亿，“一省一报（端）”用户超过 2 亿，涌现出一批千万级用户的手机报新媒体。其中，四川手机报用户由 2013 年的 1 300 万增至 6 800 万[5]。

4 深度融合 构筑媒体新版图，http://politics.people.com.cn/n1/2017/0105/c1001-28999350.html.

5 中央网信办：大力推动全国手机报新媒体创新发展，http://news.sohu.com/20160713/n458959821.shtml.

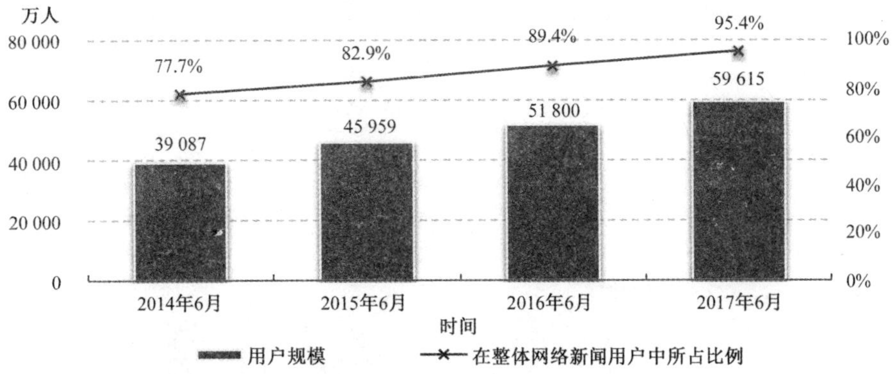

图 5-2　中国手机网络新闻用户规模和网民使用率

（数据来源：CNNIC，《中国互联网络发展状况统计报告》，2017 年 6 月）

5.1.3　互联网社交在网民生活中日益普遍

1. 即时通信开启互联网时代社交新模式

互联网即时通信与社交服务的兴起，深刻改变了中国人的沟通方式和思维方式，“无社交，不生活”成为中国网民的生活新常态。中国最早的即时通信工具是腾讯公司的 OICQ 软件，于 1999 年 3 月正式上线。此后，即时通信应用开始逐渐增多，功能不断丰富。随着微信等移动应用的快速崛起，以及移动社交网站手机版本的相继上线，在线社交逐渐从 PC 端转移至移动端。目前，我国社交网络类应用已形成完整的市场体系，涵盖 QQ、微信等即时通信工具、新浪微博等综合类社交应用，以及领域更为细分的垂直类社交应用，并且不断向泛娱乐产业链、社会化营销产业链、互联网金融产业链、大数据产业链等相关领域拓展延伸。截至 2017 年 6 月底，中国即时通信用户规模达到 6.92 亿[6]。2007—2016 年中国即时通信用户规模和使用率如图 5-3 所示，中国移动社交应用各领域代表性应用见表 5-1。

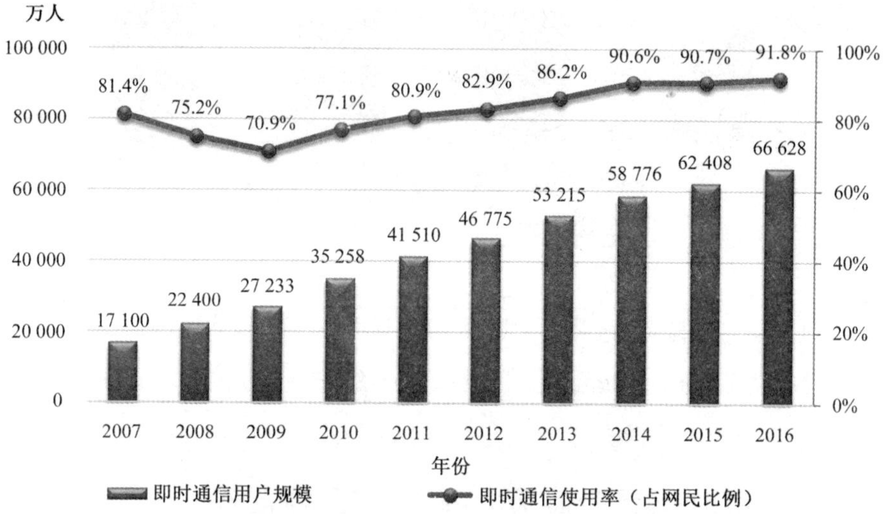

图 5-3　2007—2016 年中国即时通信用户规模和使用率

（数据来源：CNNIC，《中国互联网络发展状况统计报告》，2017 年 6 月）

表 5-1　中国移动社交应用各领域代表性应用

应用领域		代表性应用
即时通信工具		QQ、微信、飞信、易信
综合类社交应用		新浪微博、腾讯微博、QQ 空间
垂直类社交应用	社区社交	贴吧、豆瓣、知乎
	娱乐社交	美图秀秀、美拍、YY、手机对对碰
	匿名社交	无秘、平行世界、陌陌
	婚恋社交	世纪佳缘、有缘网
	职场社交	领英、脉脉

6 数据来源：CNNIC。

2．社交网络正发展为“连接一切”的生态平台

社交应用层出不穷，功能日益丰富，从即时沟通到新闻推送、视频直播、支付交易、游戏、公共服务等，都可以在社交应用上实现，覆盖多领域的平台化发展趋势明显，用户黏性不断增强。以微信为例，除了即时通信功能，微信公众号、钱包、小程序等功能先后上线，形成完整生态体系；截至 2016 年，微信支付月活跃账户数、日均支付交易笔数，都已超过 6 亿[7]。社交网络还加速了互联网商业模式多元发展，基于社交的营销服务和移动广告成为最活跃的领域，与社交圈、位置服务等功能相结合，网络营销更加精准化、个性化，成为电子商务新的增长点。

专栏 16　微信：连接一切

2011 年 1 月 21 日，腾讯推出移动即时通信——微信，成为移动互联网时代即时通信工具的代表。微信系统完全开放，支持跨通信运营商、跨操作系统、跨平台的信息传播，具备语音、短信、视频、音频、图片和文字功能，不到一年时间用户破亿。2012 年 4 月，腾讯推出微信 4.0 海外版——WeChat，支持多语言交流，成为中国模式出海的成功经验。截至 2017 年第三季度，微信月活跃用户数达到 9.8 亿，其中有 1 亿多为海外用户。在月活用户中，超过 1/3 的人每天在微信上花费 4 个多小时，微信 APP 人均每天的使用时长为 63 分钟。微信发展速度超越 QQ，遥遥领先米聊、陌陌、易信等同时期的移动即时通信工具，成为我国第一大移动端用户社交工具。经过近几年的发展，以微信为代表的即时通信工具突破用户沟通的范畴，逐渐整合线下服务，发展成为集社交、资讯、商务、服务和娱乐于一体的移动网络平台，以 IM 为入口构建生态体系，全面介入用户生活各个领域，打造在线社会生活的新模式。

7 数据来源：腾讯公司 2016 年年度业绩报告。

3. 社交媒体舆论影响力显著提升

社交媒体已成为互联网媒体中最为流行的媒体类型之一，凭借用户基数大、信息传播快、互动功能强等特点，成为影响网上舆论的重要力量。一方面，传统媒体积极拥抱社交网络，开通官方微博、微信公众号来发布权威信息，扩大传播范围，增强舆论声势。另一方面，自媒体影响力逐渐放大，截至 2017 年 6 月，微信公众平台汇集超过 2 800 万个微信公众号，其中政务类公众号达 51.4 万个，新闻类达 25.1 万个[8]，相关信息内容获得“10 万+”阅读量已成为常态。截至 2017 年第二季度，新浪微博月活跃用户达到 3.61 亿[9]，每十分钟更新一次的热门话题深度影响网络舆论热点，微博实名认证用户、网络红人等对网络话题的影响力较大。社交网络在方便人们信息交流的同时，也带来有害信息扩散、隐私泄露、过度依赖等问题，社会动员功能日益增强，给互联网管理带来重大挑战。

5.2 互联网法治体系建设日益完善

习近平总书记强调，网络空间不是“法外之地”。网络空间是虚拟的，但运用网络空间的主体是现实的，大家都应该遵守法律，明确各方权利义务。党的十八大以来，中国持续加快互联网领域立法进程，制定出台《网络安全法》这一网信领域基础性法律，颁布一批涉及网信制度的法律法规，及时修订完善互联网新闻信息服务等方面的规章规范性文件，出台互联网领域司法

8 数据来源：CNNIC 第 39 次《中国互联网络统计发展报告》。

9 数据来源：CNNIC 第 39 次《中国互联网络统计发展报告》。

解释，使现行法律延伸适用于网络空间，为互联网健康有序发展提供坚实的法治保障。

5.2.1 立法进程明显提速

1.《网络安全法》成为网信领域基本法

2016年11月7日，全国人大常委会会议表决通过《网络安全法》，并于2017年6月1日起正式施行。《网络安全法》是中国第一部全面规范网络空间安全关系的基础性法律，在中国网络空间法治建设进程中具有里程碑意义。该法首次在法律层面明确了网络产品和服务提供者的安全义务，明确了网络运营者安全义务，进一步完善了个人信息保护规则，建立了关键信息基础设施安全保护、个人信息和重要数据跨境传输等国家网络安全领域全局性、基础性的基本制度，将成为依法治网、化解网络风险的法律重器，是让互联网在法治轨道上健康运行的重要保障。

2. 相关领域立法进程不断推进

2004年8月，中国颁布《电子签名法》，确立了电子签名的法律效力。2014—2015年，中国新修订的《消费者权益保护法》《食品安全法》《广告法》等法律分别对网络销售、网购食品、网络广告等网络服务进行规范。《中华人民共和国刑法修正案（九）》将网络传谣、泄露个人信息等行为列入刑法，提高了依法管网的威慑力。2017年11月，全国人大常委会通过新修订的《反不正当竞争法》，针对电子商务领域刷单、炒作信用等不正当竞争行为进行了专门规定。

3. 两高司法解释推动法律延伸适用网络空间

为适应互联网发展治理的需求与特点，最高人民法院、最高人民检察院先后出台《最高人民法院关于审理涉及计算机网络域名民事纠纷案件适用法律若干问题的解释》《最高人民法院、最高人民检察院关于办理利用互联网、移动通讯终端、声讯台制作、复制、出版、贩卖、传播淫秽电子信息刑事案件具体应用法律若干问题的解释》《最高人民法院、最高人民检察院关于办理危害计算机信息系统安全刑事案件应用法律若干问题的解释》《最高人民法院关于审理侵害信息网络传播权民事纠纷案件适用法律若干问题的解释》《最高人民法院、最高人民检察院关于办理利用信息网络实施诽谤等刑事案件适用法律若干问题的解释》《最高人民法院关于审理利用信息网络侵害人身权益民事纠纷案件适用法律若干问题的规定》《最高人民法院、最高人民检察院关于办理电信网络诈骗等刑事案件适用法律若干问题的意见》等一系列与互联网密切关联的司法解释，推动现行法律延伸适用于网络空间，进一步推进我国互联网法治体系的完善健全。

5.2.2 政策规范相继出台

在我国互联网发展的早期阶段，互联网“接入”管理成为政策规范制订的核心，主要涉及域名管理、国际联网等议题。1996 年 2 月 1 日，国务院发布《计算机信息网络国际联网管理暂行规定》，这是在中国接入国际互联网后首次较全面地从法律层面出台规范性文件。1997 年 6 月 3 日，中国互联网络信息中心（CNNIC）成立，《中国互联网络域名注册暂行管理办法》和《中国互联网络域名注册实施细则》发布，标志着中国开始从机构、法规两方面加强网络域名管理。1998 年，公安部、信息产业部、文化部、国家工商行政管理局四部委出台《关于规范“网吧”经营行为加强安全管理的通知》，规定了网吧经营者的准入条件和审批程序。在这一阶段，网络信息安全开始纳入政策议题，国务院、公安部、人事部先后发布相关办法与条例，为网络安全管理打下基础。

2000 年之后，规范互联网应用成为政策规范制定的重点，包括网络新闻传播、网络出版、网络知识产权、网络游戏、音/视频传播、医疗卫生、电子商务、电子支付、网络税务、网络民事纠纷等领域。2000 年，国务院颁布《中华人民共和国电信条例》《互联网信息服务管理办法》，对电信市场秩序和互联网信息服务活动加以规范。2002 年 9 月 29 日，国务院颁布《互联网上网服务营业场所管理条例》，详细规定了互联网上网服务经营场所经营者的条件、义务、禁止行为与处罚制度。2004—2006 年，《互联网站禁止传播淫秽、色情等不良信息自律规范》《互联网等信息网络传播视听节目管理办法》《互联网药品信息服务管理办法》《非经营性互联网信息服务备案管理办法》《互联网 IP 地址备案管理办》《电子认证服务管理办法》《互联网著作权行政保护办法》《信息网络传播权保护条例》《互联网新闻信息服务管理规定》等一系列文件先后出台，形成互联网应用领域的基础规范体系。2007 年 6 月 25 日，《电子商务发展“十一五”规划》发布，成为中国首部电子商务发展规划，电子商务服务也率先从典型的互联网应用逐步成长为国家重要的新兴产业。

为推动互联网健康有序发展，中国坚持鼓励支持和规范发展并行，坚持政策引导和依法管理并举，先后出台一系列规范性文件，对网络购物、社交媒体、互联网金融、互联网广告、视频直播、网约车等互联网新技术新业态新模式加强规范管理。2017 年 1 月，中办、国办印发《关于促进移动互联网健康有序发展的意见》，对全方位推进移动互联网健康有序发展做出部署。国家网信办先后出台《即时通信工具公众信息服务发展管理暂行规定》《互联网用户账号名称管理规定》《互联网新闻信息服务单位约谈工作规定》《互联网新闻信息服务管理规定》《互联网信息搜索服务管理规定》《移动互联网应用程序信息服务管理规定》《互联网直播服务管理规定》《互联网跟帖评论服务管理规定》《互联网论坛社区服务管理规定》《互联网群组信息服务管理规定》《互联网用户公众账号信息服务管理规定》《互联网新闻信息服务新技术新应用安全评估管理规定》《互联网新闻信息服务单位内容管理从业人员管理办法》等一系列网络管理规定，商务部、交通运输部、国家网信办、工商总局、“一行三会”等部门和行业协会也出台了《关于促进网络购物健康发展的指导意见》《网络预约出租汽车经营服务管理暂行办法》《互联网广告管理暂行办法》《关于促进互联网金融健康发展的指导意见》等一批规范性文件，及时填补互联网新兴领域立法空白，在规范管理中推动互联网健康发展。

5.2.3 网络执法规范化、常态化

1. 建立健全网络内容管理执法体系

2013 年以来，全国网信执法三级体系建设不断推进，体制机制不断完善，建立工信、公安、工商等 12 个涉网部门之间行政执法协调工作机制，推进约谈督促警示机制建设。全面推进网信执法规范化建设，组织开展全国网信执法专题培训，统一核发新的互联网信息内容管理行政执法证。2017 年 5 月，国家网信办出台《互联网信息内容管理行政执法程序规定》，统一执法证据标准，规范执法行为。2017 年 9 月，北京、广东网信办就平台存在用户传播暴力恐怖、虚假信息、淫秽色情等信息问题，分别对新浪微博、百度和腾讯作出行政处罚。

2. 持续打击网络违法犯罪活动

针对网络诈骗、个人信息侵害等网络违法犯罪活动，中国不断加大执法力度，组织开展多项联合整治行动，取得明显成效。坚决依法查处典型案件，“魏则西事件”联合调查组要求百度对付费搜索竞价排名机制作出全面整改，“徐玉玉案”电信诈骗团伙全部落网，快手平台“伪公益”、“罗一笑”事件涉事款项全部返还网民。网信、工信、公安、工商等部门联合开展搜索引擎网站专项治理、互联网广告专项治理，严厉打击个人信息泄露、电信诈骗等违法犯罪活动。自 2017 年 3 月公安部开展打击整治黑客攻击破坏和网络侵犯公民个人信息犯罪专项行动以来，截至 7 月，全国共侦破侵犯个人信息案件和黑客攻击破坏案件 1 800 余起，抓获犯罪嫌疑人 4 800 余名，查获各类公民个人信息 500 余亿条[10]。

3. 集中整治有害信息

针对网络谣言、涉恐涉暴、涉民族宗教等违法违规有害信息，网信、公安、工信等部门密切协作，开展治理“网络敲诈和有偿删帖”、打击整治伪基站设备违法犯罪、“婚恋网站严重违规失信”，以及“净网”、“护苗”、“剑网”等系列专项行动，净化网络空间生态。2015 年 6 月，公安部部署全国公安机关普遍建立网警常态化公开巡查执法机制，截至 2016 年年底，全国 371 家网警巡查执法单位共发现处置各类违法有害信息 110 万余条[11]。2017 年 8 月，全国“扫黄打非”办公室等中央八部门联合部署开展为期 40 天的互联网低俗色情信息专项整治行动，截至 9 月，共处置、删除色情低俗等有

10 数据来源：公安部官方网站，《打击整治黑客攻击破坏和网络侵犯公民个人信息犯罪专项行动取得阶段性成果》，2017 年 3 月 10 日。

11 数据来源：公安部官方网站，《“公安改革两年间”：网络社会治理能力持续提升》，2017 年 2 月 13 日。

害信息超过 2 000 万条[12]。2016 年全国网信系统全年依法约谈违法网站 678 家，会同工信部门取消违法网站许可或备案、关闭违法网站 3 467 家[13]。

5.2.4 互联网司法创新推进

中国在稳步推进依法管网的进程中，积极探索利用互联网新技术新应用，创新互联网管理理念思路、手段举措等，使管理创新紧跟互联网创新发展，形成诸多有益实践，切实提升了管理有效性。最高人民法院试点建设“电子法院”，截至 2016 年 11 月，全国 3 520 个法院、9 277 个人民法庭和海事法庭全部接入法院专网[14]，全国法院实现一张网办案、办公，同时探索将诉讼由“线下”搬到“线上”，有效缓解了“案多人少”的矛盾。2015 年 3 月，中国首个仲裁服务网络平台——广州仲裁委员会服务律师平台在广州成立，平台提供在线立案、案件查询与统计、裁判文书下周等 30 多项功能，打破地域限制，足不出户开展仲裁业务。2015 年 8 月，浙江法院电子商务网上法庭正式上线，截至 2017 年 6 月中旬，共受理各类案件申请 22 787 件、开庭平均用时 0.5 小时[15]。2017 年 8 月 18 日，杭州互联网法院挂牌成立，成为全国第一家集中审理涉网案件的试点法院，也成为深入贯彻“网上案件网上审”的审理思维、高度融合互联网技术与机制创新、专业构建涉网纠纷解决体系的典型探索。

---

12 数据来源：全国“扫黄打非”工作小组办公室官方网站，《“大力清查互联网低俗色情信息 专项整治行动阶段性成效显著”》，2017 年 9 月 25 日。

13 数据来源：中国网信网，《严执法形成震慑 抓规范固本强基——2016 年全国网信行政执法工作扎实推进》，2017 年 1 月 20 日。

14 数据来源：《法治蓝皮书 • 中国法院信息化发展报告 No.1（2017）》。

15 数据来源：法制网记者 王春，杭州互联网法院的前世今生，法制日报——法制网（2017 年 6 月 29 日）。

专栏 17　全国首家互联网法院在杭州成立

2017 年 6 月 26 日，中央全面深化改革领导小组审议通过了《关于设立杭州互联网法院的方案》；8 月 8 日，最高人民法院明确将依托杭州铁路运输法院，试点设立专门审理涉互联网案件的杭州互联网法院。2017 年 8 月 18 日，分别在杭州和北京的原被告双方，通过互联网参与了一场“线上甄嬛告网易”的庭审，从开庭到同意调解，仅用了 20 分钟。就在这一天，全国首家互联网法院——杭州互联网法院正式揭牌成立。法院集中管辖杭州市辖区内基层人民法院有管辖权的涉互联网一审案件，在这里，起诉、立案、送达、举证、开庭、裁判，每个环节全流程在线，诉讼参与人的任何诉讼步骤即时连续地进行记录留痕，当事人能够“零在途时间、零差旅费用支出”完成诉讼[16]。

杭州互联网法院的诞生，标志着虚拟空间法治范式的重大变革，实现了“治理互联网”与“利用互联网治理”的真正融合。法院利用大数据分析技术对涉网案件数据进行多模块比对分析，梳理规律和特点，形成了结构化、标准化的互联网司法裁判规则，构建了前置性指导化解、在线争端解决方法（ODR）、第三方调解、诉讼等多层次、多元化的涉网纠纷解决体系，营造更加安全、公正、高效并且人性化的网络空间司法体系。

5.3　互联网多元治理生态完整构建

中国政府综合运用法律规制、行政管理、产业政策、技术标准、宣传教

16 人民日报记者徐隽，《全国首家互联网法院落户杭州“键对键”打官司》，《人民日报》法治头条（2017 年 9 月 6 日）。

育等措施，统筹协调各方积极参与互联网治理，引领调动互联网企业、社会组织、专家学者、智库和全体网民等各方力量，共同推进互联网多元治理生态的发展，建立起既符合互联网规律又具有中国特色的互联网治理模式。

5.3.1 企业主体作用有效发挥

1. 严格落实企业主体责任

中国互联网企业积极响应政府互联网治理的各项举措，主动履行企业主体责任。2013 年，百度、奇虎 360 等搜索引擎服务企业先后发布“网民权益保障计划”，针对搜索结果中的恶意欺诈信息和钓鱼网站给用户造成损失的予以赔偿。按照企业主体责任要求，互联网企业制定完善管理制度，及时处置微博、博客、微信、论坛、贴吧上的各类违法违规账号和有害信息。2016 年，百度网盘、新浪网盘、UC 网盘等网盘服务商落实国家版权局要求，对侵权盗版等有害信息进行全面清理。中国电信、中国移动、中国联通等电信运营企业加强治理技术提升和制度机制建设，主动堵截“虚假来电”、推出防欺诈短信提醒、全面打击“伪基站”诈骗等有害信息。同时，互联网企业依法依规维护网络秩序，以腾讯游戏“防沉迷系统”、安卓渠道联运网游的“实名认证”等为代表的企业在网络服务产品设计初期，就考虑治理重点与难点，提前布局自律机制。

2. 积极建立行业自律组织

互联网企业在不同领域、不同层面纷纷建立行业自律同盟、开展行业自律行动。2004 年 11 月，新浪、搜狐、网易公布中国互联网行业“诚信自律同盟”的自律细则，中国互联网信息服务企业主动承担维护市场秩序的主体责任。2012 年 11 月，百度、即刻、盘古、奇虎 360、盛大文学、搜狗等 12

家互联网企业在北京签署“互联网搜索引擎服务自律公约”，共同倡议坚决抵制违法有害信息传播，遵循行业惯例与商业规则，营造良性竞争环境，保护知识产权，保护用户隐私和个人信息安全等。2013 年 6 月，阿里巴巴、腾讯、百度、新浪、盛大、网易等 21 家互联网企业成立了“互联网反欺诈委员会”，推进全网联合打击网络诈骗，形成了电子商务产业生态圈联防联打的战略合作框架。

5.3.2　网络社会组织发挥积极作用

2001 年 5 月，中国互联网协会成立，先后发布《中国互联网行业自律公约》《中国互联网协会反垃圾邮件规范》《中国互联网网络版权自律公约》等公约倡议，设立了反垃圾邮件协调小组、行业自律工作委员会、网络版权联盟等机构，促进互联网行业规范有序发展。2011 年 8 月，中国网络视听节目服务协会成立，先后发布《中国网络视听节目服务协会自律公约》《关于开展“讲文明树新风”网络视听公益广告传播的倡议书》等，共同抵制不良内容、打击盗版、倡导版权保护。2012 年 12 月，中国互联网上网服务营业场所行业协会成立，有力规范营业场所经营秩序、保障经营者和网民的合法权益、促进上网服务行业转型升级。2015 年 4 月，中国文化网络传播研究会成立，促进中华优秀文化在互联网时代的创新发展，推动中华优秀文化走向世界。2015 年 8 月，全球首家互联网公募基金会“中国互联网发展基金会”在北京挂牌，积极整合社会资源、调动社会力量，资助互联网事业健康发展。2016 年 3 月，中国网络空间安全协会和中国互联网金融协会同时成立，分别在发动社会力量参与维护网络安全、引导互联网金融行业规范发展方面发挥积极作用。近年来，首都互联网协会等一批地方性网络社会组织相继成立，组织发起了多个倡导网络文明的活动，促进了良好网络素养的形成。2017 年 6 月 7 日，首次全国网络社会组织工作推进会动员部署了各地网络社会组织“同心圆”工程，公布了 100 个 2016—2017 年度（第一批）全国网络社会组织品牌项目。截至 2016 年 12 月，全国网络社会组织的总量扩大到 1 333 家，与 2015 年相比增长 144%；其中，全国性网络社会组织 47 家，地方性网络社会组织 1 286 家[17]，浙江、江苏、四川、广东四省的网络社会组织建设目前领先，数量均超过 100 家。

5.3.3 网民网络素养持续提升

1. 公众参与互联网治理积极性不断提升

中国充分调动广大公众参与互联网治理，有效维护网民自身权益，推动净化网络空间，切实维护网络空间秩序。公众监督已经成为互联网治理多元生态不可或缺的重要组成部分。自 2004 年以来，中国先后成立互联网违法和不良信息举报中心、网络违法犯罪举报网站、12321 网络不良与垃圾信息举报受理中心、12390 扫黄打非新闻出版版权联合举报中心等公众举报受理机构，不断完善违法违规信息举报工作机制，鼓励公众积极参与互联网监督和治理。2016 年，全国共受理网络举报近 4 000 万件次，有害信息有效处置率达 92%以上[18]。

2015 年，中国开展“中国好网民”系列活动，2016 年初启动争做中国好网民工程，积极推进网民网络素养教育，推出网络大课堂、公益广告设计大赛等系列主题活动，连续两届在国家网络安全宣传周期间举办网络素养主题论坛，引导社会各方共同参与网民网络素养教育工作，让遵纪守法、理性

17 数据来源：首次全国网络社会组织工作推进会发布数据（2017 年 6 月 7 日）。

18 数据来源：中国互联网违法和不良信息举报中心。

表达、文明上网的中国好网民理念深入人心。

2. 民间网络监督团体不断涌现

中国网民在主动抵制网上非法行为、主动参与网上文明活动的同时，自发构建起不同类型的民间网络监督团体，积极贡献文明网络、绿色网络、和谐网络建设。“妈妈评审团”“互联网妈妈监督团”“辟谣联盟”等网络监督团体，在保护青少年健康上网、建立网民反馈监督机制、规范网络传播秩序等方面做出了积极努力，承担起净化网络环境、构建舆论生态的重要职责。

专栏18　中国网民自发成立网络监督团体

“妈妈评审团”发起人是北京市青少年法律与心理咨询服务中心主任宗春山，成员主要通过招募方式由未成年人家长组成。其基本职能是依据“儿童利益最大原则”，由“妈妈们”对互联网上影响未成年人身心健康的内容进行举报、评审，形成处置建议反映给相关管理部门，并监督评审结果的执行，为社会公众参与网络不良信息的监督评审搭建了一座新的桥梁。

“辟谣联盟”成立于2011年5月18日，由窦含章发起，是网络上首个民间辟谣组织。在新浪、微博账号“辟谣联盟”背后，活跃着一群热心于澄清事实、还原真相的网民，成员达到数千人；在自媒体时代复杂的网络环境下，辟谣联盟一方面成为网民自我约束的自律组织，另一方面又承担着净化微博环境的重要职责，成为网民参与互联网治理的创新性探索。

第 6 章　网络安全保障能力显著提升

没有网络安全就没有国家安全，没有信息化就没有现代化。网络空间已成为继陆、海、空、天之后的第五维空间，是大国博弈的焦点所在，也是维护国家安全的战略新疆域。中国以习近平总书记提出的“树立正确的网络安全观”为指引，以安全保发展、以发展促安全，坚持网络安全管理和技术并重，建立完善的网络安全审查机制，强化关键信息基础设施安全保护，不断提高广大人民群众网络安全防护意识和技能，切实构筑起坚实的网络空间安全屏障。

6.1　网络安全总体态势严峻

新一代信息技术在各领域快速渗透，网络安全环境日趋复杂，网络安全风险进一步加大，关键信息基础设施、个人信息安全、移动互联网等领域的安全事件时有发生，高级持续性威胁（APT）、分布式拒绝服务（DDoS）、木马和僵尸网络、勒索病毒等网络攻击持续增加，我国网络安全形势依然严峻。

6.1.1　各领域网络安全风险凸显

1. 网站安全问题频发

近几年，网页篡改、网站后门、网页仿冒等问题频发，尤其在 2014 年

Windows XP 系统停止更新服务后，国内大量使用 Windows XP 操作系统的用户面临不少现实与潜在的安全风险。国家互联网应急中心（CNCERT）数据显示，2017 年前三季度，中国境内被篡改网站数量 17 353 个，其中政府网站数量 543 个；境内 24 353 个网站被植入后门，其中政府网站 1 191 个；仿冒我国境内网站的钓鱼页面 16 099 个，涉及 IP 地址 4 026 个，比 2016 年大幅下降。但是，基础应用或通过软硬件漏洞的威胁严重，2017 年 3 月发现 Apache Struts2 S2-045 远程代码执行漏洞，攻击者利用该漏洞可直接取得网站服务器控制权，采用该框架的网站受影响比例达 60.1%。2013—2016 年网站安全监测情况如图 6-1 所示。

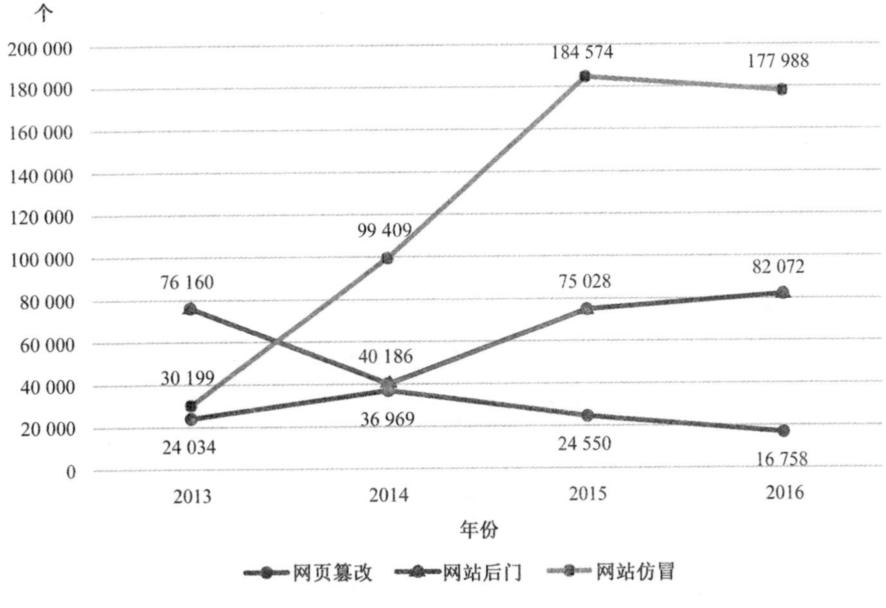

图 6-1　2013—2016 年网站安全监测情况

（数据来源：CNCERT）

专栏 19　网站攻击手段

网页篡改是恶意破坏或更改网页内容，使网站无法正常工作或出现黑客插入的非正常网页内容。根据攻击手段，可以分成显式篡改和隐式篡改两种。通过显式网页篡改，黑客可炫耀自己的技术技巧，或达到声明自己主张的目的。隐式篡改一般是在被攻击网站的网页中植入被链接到色情、诈骗等非法信息的暗链中，以助黑客谋取非法经济利益。黑客为了篡改网页，一般需提前知晓网站的漏洞，提前在网页中植入后门，并最终获取网站的控制权。

网站后门是指黑客在网站的特定目录中上传远程控制页面，从而能够通过该页面秘密远程控制网站服务器的攻击形式。通过在网站的特定目录中上传远程控制页面，黑客可以暗中对网站服务器进行远程控制，上传、查看、修改、删除网站服务器上的文件，读取并修改网站数据库的数据，甚至可以直接在网站服务器上运行系统命令。

网页仿冒俗称网络钓鱼（Phishing），是社会工程学欺骗原理与网络技术相结合的典型应用，是通过构造与某一目标网站高度相似的页面诱骗用户的攻击方式。钓鱼网站是网页仿冒的一种常见形式，常以垃圾邮件、即时聊天、手机短信或网页虚假广告等方式传播，用户访问钓鱼网站后可能泄露账号、密码等个人隐私。

2. 关键信息基础设施安全风险增大

随着“互联网+”、智能制造等的快速发展，互联网快速渗透到社会生产生活各领域各环节，关键基础设施特别是重要行业信息系统或工业控制系统，逐渐成为网络攻击的重灾区，一旦发生网络安全事故，将对国家和人民造成严重损害。截至 2017 年 9 月底，国家信息安全漏洞共享平台（CNVD）共收录工业控制漏洞 1 313 条，其中 2017 年收录 307 条，比 2016 年增长 77.5%；

主要工业控制漏洞风险包括缓冲区溢出、缺乏访问控制机制、弱口令、目录遍历等[1]。CNCERT 共发现我国联网工业控制设备 72 767 个，协议主要涉及 S7Comm、Modbus、SNMP、EtherNetIP、Fox、FINS、BACnet 等，厂商主要为西门子、罗克韦尔、施耐德、欧姆龙、风河、韦益可自控等，以国外产品为主，我国工业互联网的安全可控之路任重道远；累计监测到联网工业控制设备指纹探测事件 820 万余次，并发现来自境外 160 个国家的 46.5 万余个 IP 地址对我国联网工业控制设备进行指纹探测，比 2016 年呈爆发式增长。

3. 信息泄露事件屡见不鲜

随着互联网不断跨界发展，数据和信息遍布网络、手机和云上，受到互联网黑色产业链的利益驱动，网站数据和个人信息泄露事件屡见不鲜，规模和范围不断扩大，严重威胁个人财产乃至人身安全。从 2014 年年底开始，中国连续出现各类信息泄露事件。2014 年 12 月，铁道部官方网站（12306.cn）13 万用户身份证、登录口令等信息泄露，2015 年 2 月桔子、锦江之星等知名连锁酒店，以及万豪、喜来登等高端酒店客户订单信息泄露，2015 年 8 月华语第一票务网站大麦网 600 多万用户账号密码泄露，2015 年 10 月网易邮箱过亿用户敏感信息遭泄露，2016 年 4 月济南 20 万儿童信息被公开打包出售，2016 年 8 月高考考生信息泄露间接夺去“准大学生”徐玉玉的生命，2017 年 7 月百度网盘用户公开分享资料被泄露。

1 数据来源：CNCERT。

专栏 20　徐玉玉被电信诈骗致死案

2016 年 8 月 19 日，山东临沂一名家境贫寒的 18 岁“准大学生”徐玉玉，接到冒充教育局、财政局工作人员“发放 2 600 元助学金”的诈骗电话，被骗光了父母辛苦借来的 9 900 元学费。发现被骗后，徐玉玉伤心欲绝，在报警回家途中突发心脏骤停，经医院抢救无效，不幸离世。此案件引起全社会广泛关注。

接到报警之后，警方高度重视，成立专案组，全力侦破此案。2017 年 7 月 19 日，“徐玉玉案”一审公开宣判，主犯陈文辉因诈骗罪、非法获取公民个人信息罪被判无期徒刑，没收个人全部财产；其他六名被告人被判 15 年到 3 年不等的有期徒刑并处罚金。9 月 15 日，“徐玉玉案”二审裁定驳回上诉维持原判。此外，在“徐玉玉案”关联案件的审判中，“黑客”杜天禹也因非法获取公民个人信息罪被判有期徒刑 6 年，并处罚金 6 万元。

4．域名系统安全状况有所改善

域名系统一度是影响我国互联网稳定运行的薄弱环节，域名系统面临严峻的拒绝服务攻击，针对重要网站的域名解析篡改攻击频发。2014 年国内出现大规模的 DNS 故障，全国众多门户网站无法打开，根据 IP 定位等多种证据显示，这几轮大规模 DNS 攻击均来自境外。随着我国重要域名系统安全防护措施的普遍加强，我国域名服务系统安全状况已明显好转。近两年，针对重要域名系统的 DDoS 攻击事件均未对相关系统的域名解析服务造成严重影响，在基础电信企业侧也未发生严重影响解析成功率的攻击事件。2016 年 6 月，发生针对全球根域名服务器及其镜像的大规模 DDoS 攻击，因应急处置及时，未对我国域名系统网络安全造成影响。

专栏21　2014年全国发生大面积DNS域名解析故障事件

2014年1月21日15时10分左右，国内通用顶级域的根服务器解析出现异常，部分国内用户无法访问.com的域名网站。百度、新浪、腾讯、京东、优酷等大批网站出现间歇性无法访问的情况，网站被解析到了65.49.2.178这一个无法访问的美国IP地址。据初步统计，此次全国有2/3的网站访问受到影响。截至当日16点，全国根服务器的解析陆续恢复正常。

5. 移动互联网安全问题日益突出

随着移动互联网使用的不断增多，针对移动端的攻击愈发突出，移动互联网安全防御成为新的工作重点。2017年前三季度，通过自主监测和共享方式累计捕获移动互联网恶意程序近168万个，近7年持续保持高速增长趋势；其中，数量排名前两位的分别是流氓行为类（54.5%）和资费消耗类（37.7%）。同时，2017年前三季度，移动互联网恶意程序下载链接近146万条，较2016年全年增长近1.2倍，恶意程序传播次数达0.2亿次。在恶意APP监测方面，连续5年治理其传播途径得到有效控制，到2017年9月，CNCERT累计向141家已备案的应用商店、网盘、云盘的广告宣传等网站运营者通报恶意APP事件8 166起。值得注意的是，移动支付安全问题不容小觑，社交账号盗用、短信木马链接、骗取验证码成为最常见的诈骗手段[2]。2010—2016年移动互联网恶意程序样本数量对比如图6-2所示。

---

2 数据来源：中国银联，《2016移动支付安全大调查分析报告》。

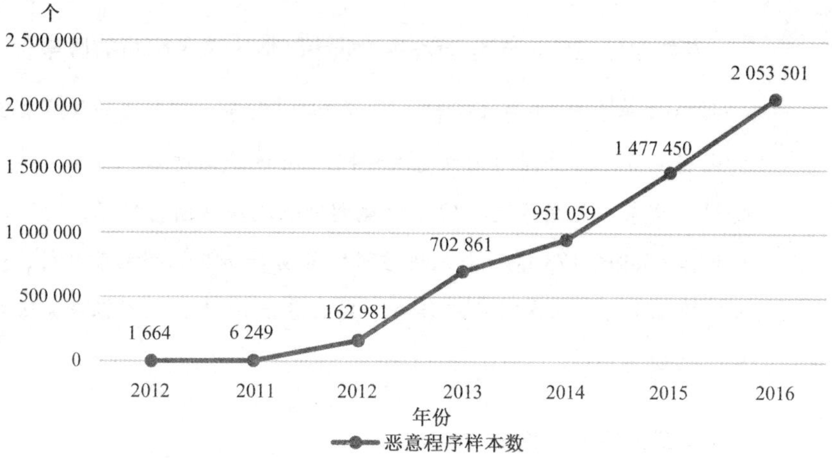

图 6-2　2010—2016 年移动互联网恶意程序样本数量对比

（数据来源：CNCERT）

6. 物联网智能设备安全问题不断出现

智能可穿戴设备、智能家居、智能路由器等终端设备和网络设备正迅速成为人们工作和生活的重要组成部分，针对物联网智能设备的网络攻击事件比例呈现上升趋势。攻击者利用物联网智能设备漏洞可获取设备控制权限，进而被控制形成大规模僵尸网络，或者窃取用户信息数据、劫持网络流量等进入黑客地下交易市场。2017 年前三个季度，CNVD 共收录物联网智能设备漏洞 531 个，主要涉及网络摄像头、智能路由器、智能家电、智能网关等设备；漏洞类型主要为权限绕过、信息泄露、命令执行等。其中弱口令（或内置默认口令）漏洞极易被利用，成为恶意代码攻击利用的重要风险点[3]。随着无人机、自动驾驶汽车、智能家电的普及和智慧城市的

3 数据来源：国家互联网应急中心（CNCERT）。

发展，未来或利用物联网智能设备的网络攻击将更为频繁。

7. 云端安全事件大量增加

随着云端业务和数据的逐步累积，针对云端的基于漏洞、病毒、未知威胁的 APT 攻击、0Day 攻击等日益增加，云安全事件频频发生。2014 年年底，部署在阿里云上的一家知名游戏公司，遭遇了当时全球最大的一次 DDoS 攻击，攻击时间长达 14 个小时，攻击峰值流量达到 453.8Gbps[4]。2015 年上半年，腾讯云有数千台主机遭受外部黑客 1 万多次 DDoS 攻击，峰值月份受到攻击次数达到 2 600 多次；被攻击的最大流量也逐月上涨，最高峰值月份接近 300Gbps[5]。云安全联盟总结了 2016 年“十二大云安全威胁”，分别为数据泄露、凭证被盗和身份验证如同虚设、界面和 API 被黑、系统漏洞利用、账户劫持、恶意内部人士、APT 寄生虫、永久的数据丢失、调查不足、云服务滥用、拒绝服务（DoS）攻击、共享技术等[6]。

6.1.2 网络攻击手段层出不穷

1. 中国正成为 APT 攻击主要受害国

当前，日益频繁的 APT 攻击正在导致政企行业机密情报被窃取、工业系统被破坏、金融系统遭受经济损失，严重地威胁着国家安全和公民权益。360 威胁情报中心累计监测数据显示，截至 2016 年 12 月底，监测到针对中国境内目标发动攻击的境内外 APT 组织 36 个，疑似 APT 攻击目标的境内组

---

4 数据来源：阿里云关于阿里云游戏用户遭遇黑客攻击的声明，2014 年 12 月 24 日。

5 数据来源：腾讯云，腾讯云安全白皮书。

6 信息来源：云安全联盟在腾讯“云+未来”峰会上的主题演讲《国际云安全趋势与实践》。

织机构近 200 个，中国正成为全球 APT 攻击的第一目标国。仅 2016 年，这些 APT 组织发动的攻击行动至少影响中国境内超过上万台计算机，攻击范围遍布国内 31 个省级行政区，攻击对象包括政府、基础设施、教育、科研、大型企业等机构[7]，如“白象行动”、“蔓灵花攻击行动”等都是针对我国重要信息系统实施的 APT 攻击事件。目前，更多的联网生产系统和基础设施面临 APT 攻击的巨大网络安全威胁。

2. DDoS 攻击不断

DDoS 攻击已经成为最引人注目的、最受黑客欢迎的攻击方式。绿盟科技监测数据显示，与 2017 年上半年相比，2016 年整体攻击趋势放缓，监控到 DDoS 攻击约 10 万次，相比 2016 年下半年下降了 30%；攻击总流量约为 1.6 万 Tbps，相比 2016 年下半年下降了 38.4%，但峰值大于 300Gbps 的超大流量攻击却呈增长趋势[8]。从攻击范围来看，政府、金融、游戏行业是 DDoS 攻击的重灾区，且 DDoS 威胁的不再是个别重点行业，各行各业都将面临 DDoS 攻击的巨大威胁。此外，黑客攻击手段也越来越高明，在 DDoS 攻击的同时，还开始从事窃取数据、感染、欺骗等其他形式的网络犯罪，以套取更大利益。

3. 木马和僵尸网络感染情况有所缓解

CNCERT 抽样监测结果显示，2017 年前三季度，在利用木马或僵尸程序控制服务器对主机进行控制的事件中，受控主机 IP 地址总数为 1 662 万个。其中，境内 1 138.5 万余个 IP 地址对应的主机被木马或僵尸程序控制，境外 523.5 万余个 IP 地址对应的主机被木马或僵尸程序控制。这三类事件

---

7 数据来源：360 互联网安全中心，《2016 年中国互联网安全报告》。
8 数据来源：绿盟科技，《2017 上半年 DDoS 与 Web 应用攻击态势报告》。

分别比2016年的2 584万个、1 700万个和885万个有了大幅下降。控制服务器IP地址总数约为18.2万个，其中，境内木马或僵尸程序控制服务器IP数量为13万余个，境外木马或僵尸程序控制服务器IP数量为5.2万余个，却较2016年大幅上升[9]。

4．勒索病毒集中爆发

勒索病毒是一种新型电脑病毒，严重威胁了本地数据和智能设备的安全。2017年前三季度，根据CNCERT监测，在传统PC端捕获敲诈勒索类恶意程序样本约13.93万个。勒索病毒的攻击对象逐渐由个人终端设备延伸至企业用户，在企业用户层面，主要表现为攻击企业加密数据库；在个人终端设备层面，主要通过“加密数据”进行传统PC端勒索以及通过“加密设备”进行移动端勒索。2017年5月，“WannaCry”敲诈勒索病毒席卷全球，也感染了我国大量行业企业内网，教育、医疗、电力、能源、银行等多个行业受到影响，中石油部分加油站出现了加油卡、银行卡、第三方支付等网络支付无法使用的状况。

专栏22　攻击手段

APT（Advanced Persistent Threat），即高级持续性威胁，是黑客以窃取核心资料为目的，针对客户所发动的网络攻击和侵袭行为。APT的攻击手法，在于隐匿自己，针对特定对象，长期、有计划和有组织地窃取数据。这种发生在数字空间的偷窃资料、收集情报的行为，就是一种“网络间谍”的行为。

DDoS（Distributed Denial of Service），即分布式拒绝服务攻击，指借助客户/服务器技术，将多个计算机联合起来作为攻击平台，对一个

9 数据来源：国家互联网应急中心（CNCERT）。

或多个目标发动 DDoS 攻击，主控程序能在几秒内激活成百上千次代理程序的运行，从而成倍地提高拒绝服务攻击的威力。

僵尸网络，指被黑客集中控制的计算机群，其核心特点是黑客能够通过一对多的命令与控制信道，操控感染了木马或僵尸程序的主机执行相同的恶意行为，例如，可同时对某目标网站进行 DDoS 攻击，或者发送大量的垃圾邮件等。

勒索病毒，是一种新型计算机病毒，主要以邮件、程序木马、网页挂马的形式进行传播，利用各种加密算法对文件进行加密，被感染的用户一般无法解密，必须拿到解密的私钥才有可能破解。

6.2 网络安全体制（机制）加快建设和完善

2012 年以来，中国以网络安全战略规划为引领，以完善网络安全体制机制为突破口，以建立网络安全监测和应急处置体系为基础，进一步摸清网络安全管理现状，明晰网络安全保障重点，不断完善相关法律法规和工作机制，培养一批专业支撑机构，网络安全保障能力明显提升。

6.2.1 网络安全战略规划引领

面对日益严峻复杂的网络安全总体态势，中国坚持规划先行。党的十八大以来，中国出台了《国家网络空间安全战略》《国家信息化发展战略纲要》《“十三五”国家信息化规划》等涉及网络安全领域的多个战略规划，为网络安全工作指明方向。2016 年 7 月，《国家信息化发展战略纲要》发布，提出要树立正确的网络安全观，坚持积极防御、有效应对，增强网络安全防御能力和威慑能力，维护网络主权和国家安全，加强关键信息基础设施安全防护，强化网络安全基础性工作。2016 年 12 月，国家网信办发布《国家网络空间安全战略》，作为指导国家网络安全工作的纲领性文件，以总体国家安全观为指导，统筹发展与安全两件大事，推进网络空间和平、安全、开放、合作、有序，明确了我国网络安全的四个原则、九项战略任务。 2016 年 12 月，国务院印发《“十三五”国家信息化规划》，坚持安全与发展并重，将“健全网络安全保障体系”作为一项重要任务，提出“强化网络安全顶层设计、构建关键信息基础设施安全保障体系、全天候全方位感知网络安全态势、强化网络安全创新能力”等重大任务及工程。

6.2.2 网络安全法律法规体系不断健全

1. 网络安全相关法律逐渐建立完善

《网络安全法》于 2017 年 6 月 1 日起施行，规定了国家网络安全工作的基本原则、主要任务和指导思想、理念，明确了政府部门、企业、社会组织和个人的权利、义务和责任，是我国网络安全领域基础性、框架性、综合性法律。2015 年 7 月，《中华人民共和国国家安全法》正式实施，提出建设网络与信息安全保障体系，提升网络与信息安全保护能力，加强网络和信息技术的创新研究和开发应用，维护国家网络空间主权、安全和发展利益。《中华人民共和国刑法修正案（九）》对利用信息网络编造传播虚假险情、疫情、灾情、警情信息等网络犯罪行为做出明确规定。2016 年 1 月，《中华人民共和国反恐怖主义法》正式实施，规定了网络空间防范恐怖主义等内容。

2. 网络安全政策法规陆续出台

2003 年，《国家信息化领导小组关于加强信息安全保障工作的意见》印发，首次为网络安全工作确立了基本纲领和大政方针，提出了加强网络安全保障的指导思想和主要任务。此后，《互联网 IP 地址备案管理办法》、《信息安全等级保护管理办法》《电子认证服务密码管理办法》《关于大力推进信息化发展和切实保障信息安全的若干意见》《互联网网络安全信息通报实施办法》《通信网络安全防护管理办法》等文件陆续出台。2012 年，全国人大常务委员会出台《全国人民代表大会常务委员会关于加强网络内容信息保护的决定》。2013 年 7 月，工信部发布了《电信和互联网用户个人信息保护规定》，规范电信服务、互联网信息服务过程中收集、使用用户个人信息的活动，切实保护电信和互联网用户的合法权益。2014 年 5 月，中央网信办印发《关于加强党政机关网站安全管理的通知》，要求全面提升党政机关网站安全保障能力和管理水平，确保党政机关网站安全运行、健康发展。2014 年 8 月，工信部发布《关于加强电信和互联网行业网络安全工作的指导意见》，要求从网络基础设施安全防护、突发网络安全事件应急响应、安全可控关键软硬件应用、网络数据和用户个人信息保护等方面加强网络安全监管。2017 年 1 月，中央网信办印发《国家网络安全事件应急预案》。2017 年 8 月，工信部修订完善《互联网域名管理办法》，进一步优化 IP 地址、域名、网站备案管理。

6.2.3 网络安全工作机制逐渐完善

1. 网络安全审查制度逐渐建立完善

2014 年 5 月，国家有关部门提出将启动网络安全审查，将关系国家安全和公共利益的系统、使用的重要信息技术产品和服务纳入审查范围。2014 年 12 月，中央网信办出台《关于加强党政部门云计算服务网络安全管理的意见》，提出建立针对党政部门云计算服务提供商的云计算服务安全审查机制，审查其云计算服务的安全性、可控性。2017 年 5 月，国家网信办出台《网络产品和服务安全审查办法（试行）》，建立完善网络安全审查机制，明确关系国家安全的网络和信息系统采购的重要网络产品和服务应当经过网络安全审查。目前，浪潮软件集团有限公司所建济南政务云平台、曙光云计算技术有限公司所建成都电子政务云平台（二期）和阿里云计算有限公司所建阿里云电子政务平台等多个平台通过网络安全审查。

2．关键信息基础设施网络安全检查工作持续进行

按照习近平总书记“要全面加强网络安全检查，摸清家底，认清风险，找出漏洞，通报结果，督促整改”的要求，自 2014 年起，中央网信办就组织开展了针对中央和国家机关各部委、各人民团体，以及重点行业重要网络与信息系统的网络安全大检查。2016 年，由中央网信办牵头，各省（自治区、直辖市）网信办统筹组织，首次开展了全国范围的关键信息基础设施网络安全检查工作，形成全国检查一盘棋工作格局，以查促管、以查促防、以查促改、以查促建，为构建关键信息基础设施安全保障体系提供基础性数据和参考。此次检查共排查 31 省（自治区、直辖市）的 5.4 万余个信息系统，覆盖全国 90%的县级以上行政区域。

3．持续开展信息安全等级保护

2007 年 1 月，公安部出台《信息安全等级保护管理办法》，规定对信息和信息载体按照重要性等级实施信息安全等级保护，并逐渐探索出一系列从中央到地方的信息安全等级保护工作机制。信息安全等级保护工作包

括定级、备案、安全建设和整改、信息安全等级测评、信息安全检查五个阶段；安全等级则根据信息系统受到破坏后造成的威胁程度，划分为第一级到第五级，要求不同安全等级的信息系统应具有不同的安全保护能力。《网络安全法》规定，国家实行网络安全等级保护制度，网络运营者应当按照网络安全等级保护制度的要求，履行安全保护义务，保障网络免受干扰、破坏或者未经授权的访问，防止网络数据泄露或者被窃取、篡改。

6.2.4 网络安全保障服务机构支撑有力

1. 国家级保障服务机构不断健全

经过 20 年的发展，我国已建成国家级网络与信息安全应急技术的五大保障服务机构。一是国家计算机病毒应急处理中心（CVERC）。这是我国唯一的计算机病毒应急处理机构，建有中国互联网计算机病毒与网络攻击监测预警中心和国家移动互联网应用安全管理中心，主要负责快速发现和处置计算机病毒疫情与网络攻击事件，拥有国内最权威的恶意代码样本信息库，存储计算机病毒样本 1 500 万余个。二是中国信息安全测评中心。这是我国专门从事信息技术安全测试和风险评估的权威职能机构，建有漏洞基础研究、应用软件安全、产品安全检测、系统隐患分析和测评装备研发等多个专业性技术实验室。三是国家计算机网络应急技术处理协调中心（国家互联网应急中心，CNCERT），依托公共互联网网络安全监测平台，开展互联网网络安全事件的预防、发现、预警和协调处置等工作；目前已在全国 31 个省（自治区、直辖市）建立分中心，并联合 580 家成员单位建立了较稳定的网络安全信息通报工作体系。四是国家网络与信息安全信息通报中心。这是国家级网络和信息安全信息通报出口，对全国网络信息安全方面的信息进行报告和通报。五是国家信息技术安全研究中心，主要承担信息技术产品和系统的安全性分析与研究、国家基础信息网络和重要信息系统的信息安全保障任务以及研发具有自主知识产权的信息安全技术。六是中国信息安全认证中心，依据国家有关强制性产品认证、信息安全管理的法律法规，负责实施信息安全认证的专门机构。

2. 全国性网络安全社会组织相继成立

随着网络安全的战略地位进一步上升，相关协会、联盟等一批全国性网络安全社会组织纷纷成立。2016年3月，中国网络空间安全协会在北京成立，由国内从事网络空间安全相关产业、教育、科研、应用的机构、企业及个人共同自愿结成，现有会员单位260个、个人会员321人，囊括国内主要互联网企业和网络安全企业、权威科研机构及网络安全专家，与地方网信办联合举办“网安中国行”系列活动，提升网络安全防范意识和技能。2015年12月，中国网络安全产业联盟（筹）在北京正式成立，由蓝盾股份、绿盟科技、启明星辰、卫士通等国内140余家网络安全代表性企业自愿联合、共同发起组建。2016年1月，腾讯云、绿盟科技、启明星辰、亚信安全、IBM、赛门铁克、天融信、深信服8家知名企业联合成立云安全服务联盟。

6.3　网络安全产业与技术加速突破

网络安全已经成为需要各界共同应对的“大安全”问题，经过数十年的发展，我国网络安全自主可控技术研发取得一定突破，网络安全产业正迎来新一轮的发展机遇，网络安全相关标准正逐步制定，并着重健全网络安全人才培养体系。

6.3.1 网络安全技术取得突破

中国加强新技术研究和安全标准制定，研发和部署一系列安全产品和系统平台，攻克了一批信息安全重大技术难题。自主研发了加密强度与国际主流密码算法相当的祖冲之密码算法，安全可靠基础软硬件联合攻关取得重大进展，龙芯、申威、飞腾、海思、展讯等国产芯片加快创新突破，自主安全操作系统等已从“基本可用”迈向“可用”，国产数据库在国家机关和重要领域广泛应用，基于国产基础软硬件的国产整机在部分党政机关得到试用，工业控制系统的安全防护系统产业化取得积极进展。

1. 基础技术从空白到逐个击破

长期以来，我国在芯片、服务器、数据库、操作系统等安全可控基础技术方面长期依赖进口，十八大以来情况开始逐渐改观。2013 年，中国智能终端操作系统产业联盟成立，积极搭建统一的国产操作系统生态圈，打开市场局面。2014 年，普华、中标麒麟、思源的新版操作系统相继发布。2016 年，龙芯中科宣布 3A3000 四核处理器芯片完成流片并通过系统测试，华为发布 SOC 麒麟 960，瑞芯微 rockchip 发布高性能、高扩展应用的全能型旗舰级芯片 RK3399，飞腾公布飞腾 2000。同时，在 5nm 等离子刻蚀机、高纯度溅射金属靶材、研磨液等芯片制造关键技术取得突破，集成电路产业迎来全面爆发。2017 年我国自主研发了 10 光量子纠缠操纵的光量子计算原型机，打破世界纪录，宣告我国在量子计算领域已处于全球领先地位。

2. 安全防护技术从被动到主动防御

在国家层面，在工控安全、云安全、电子政务系统安全等领域投资建设了一批国家级信息安全研发中心，支持研发了一大批应用于云计算、物联网、移动互联网、下一代互联网等新兴领域的高性能安全网关、UTM、SOC、防火墙等安全产品和服务，有效保障电子政务、电信、金融、交通等重要领域信息系统的安全。在企业层面，2016 年，我国网络安全相关厂商不断提升自身产品性能，以满足来自各行业的网络安全需求：2 月，在著名国际独立安全研究和测评机构 NSS labs 公布的 2015 年度下一代防火墙的测试评结果中，中国厂商山石网科以 99%的综合威胁检查率和排名第一的总体拥有成本，获得“推荐奖”；4 月，匡恩网络推出首款便携式工控安全威胁评估系统平台；5 月，绿盟科技发布全新绿盟工控入侵检测系统 IDS-ICS，并获得全国首个工控入侵检测产品资质；11 月，“网络空间拟态防御理论及核心方法”通过验证，是中国在网络防御领域的重大理论和方法创新。此外，利用大数据技术进行行为分析、态势感知、追踪溯源正逐渐成为解决针对性攻击的总要手段；深度学习、人工智能、区块链、量子计算等前沿技术正在兴起，与网络安全的结合应用正在研究发展。

6.3.2　网络安全产业快速发展

1. 产业规模增长迅速

根据赛迪智库测算，2013—2016 年，网络安全产业规模从 536.5 亿元增长到 1 066.4 亿元，年均复合增速约 25%。随着政策的加速推动，网络安全产业将迎来黄金发展期，预计 2017 年，我国网络安全产业规模有望达到 1 355.4 亿元，增幅将突破 27.1%。个人计算机的安全市场已由国产安全品牌主导，国内电脑用户的免费安全软件覆盖率达到 99%以上[10]。随着网络安全产业规模的迅速扩大，涌现出一批领军企业。截至 2016 年年底，国内

10 数据来源：中国电子信息产业发展研究院，《2016—2017 年中国网络安全发展蓝皮书》。

网络安全企业超过 1 000 家，年度营收过亿的企业超过 30 家。传统网络安全企业中，涌现了启明星辰、深信服、绿盟科技、360 企业安全、天融信、卫士通、新华三等一大批国内知名的网络安全龙头企业，以及国民认证、青藤云安全、椒图科技、安全狗、威努特、梆梆安全等新兴和初创网络安全专业企业；阿里巴巴、腾讯、百度等大型互联网企业也纷纷进军网络安全产业，为网络安全产业注入新的力量。截至 2017 年 6 月，启明星辰、天融信、蓝盾股份、卫士通等 13 家企业已在 A 股主板上市，优炫软件、鼎普科技、圣博润等 38 家企业在新三板上市[11]。总体来看，尽管中国网络安全企业实力有明显提升，但与网络安全保障需求和国际网络安全企业实力相比，仍有较大差距。2013—2016 年中国主要网络安全产业规模及增长情况如图 6-3 所示；中国主要网络安全企业情况见表 6-1。

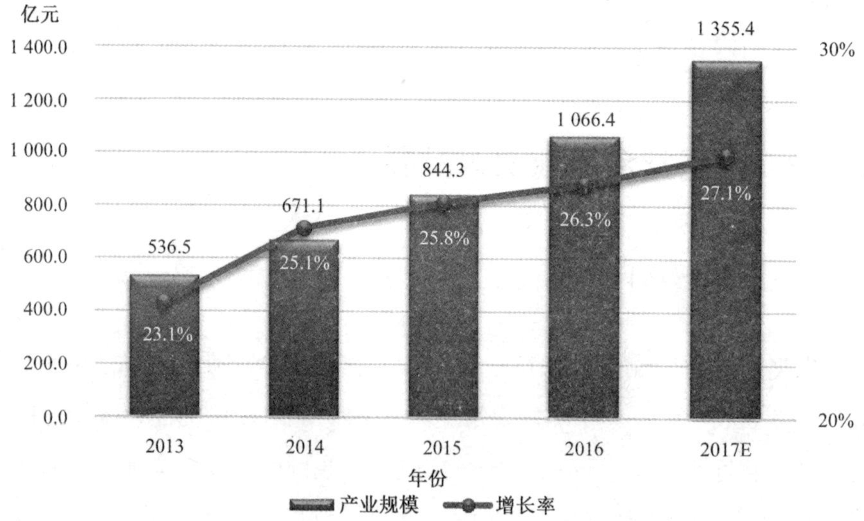

图 6-3 2013—2016 年我国网络安全产业规模及增长情况

（数据来源：赛迪智库）

11 数据来源：安全牛，《中国网络安全企业 50 强》（2017 年上半年）。

表 6-1 中国主要网络安全企业情况

序号	企业名称	主要营业领域	2016 年营业收入（亿元）	总部所在
1	启明星辰	防火墙&NGFW、网络隔离（网闸）、入侵检测（IDS）/入侵防御（IPS）、统一威胁管理（UTM）、数据库安全、数据防泄密 DLP、VPN、漏洞扫描、SOC & NGSOC、评估加固和安全运维	19.27	北京
2	深信服	防火墙和 NGFW、统一威胁管理（UTM）、上网行为管理、VPN、移动终端安全	17.5	深圳
3	绿盟科技	防火墙&NGFW、入侵检测（IDS）/入侵防御（IPS）、统一威胁管理（UTM）、主机防护及自适应、抗 DDoS 设备、数据库安全、漏洞扫描、Web 应用扫描与监控、Web 应用防火墙（WAF）、安全咨询服务、评估加固和安全运维	10.89	北京
4	360 企业安全	防火墙&NGFW、网络隔离（网闸）、统一威胁管理（UTM）、终端防护&防病毒、终端检测响应（EDR）、VPN、代码审计、Web 应用防火墙（WAF）、移动 APP 安全、移动终端安全、威胁情报、大数据安全、APT、SOC & NGSOC、渗透测试服务	—	北京
5	亚信安全	统一威胁管理（UTM）、主机防护及自适应、终端防护&防病毒、终端检测响应 EDR、防垃圾邮件、云基础架构安全、移动终端安全、APT、反钓鱼、SOC & NGSOC	—	成都
6	天融信	防火墙&NGFW、入侵检测（IDS）/入侵防御（IPS）、数据防泄密（DLP）、VPN、云基础架构安全、大数据安全	—	北京
7	卫士通	防火墙&NGFW、入侵检测（IDS）/入侵防御（IPS）、VPN、磁盘加密、文档安全、加密机	17.99	成都

续表

序号	企业名称	主要营业领域	2016 年营业收入（亿元）	总部所在
8	新华三	防火墙&NGFW、入侵检测（IDS）/入侵防御（IPS）、统一威胁管理（UTM）、VPN	260[12]	杭州
9	安恒信息	数据库安全、Web 应用扫描与监控、Web 应用防火墙（WAF）、堡垒机/运维安全、大数据安全、等保工具	—	杭州
10	美亚柏科	大数据安全、安全取证、舆情监控、安全培训教育、安全集成服务	5.92	厦门

数据来源：各企业公开年报数据。

2. 网络安全产业结构逐渐完善

网络安全产业主要包括五部分：基础安全产业、IT 安全产业、容灾备份产业、网络可信身份服务业以及其他。2016 年中国网络安全产业结构中（见图 6-4），IT 安全产业为主力，占比达到 48.4%，基础安全产业、容灾备份产业和网络可信身份服务业占比分别为 5.6%、13.1%和 32.8%。其中，IDC 统计数据显示，2016 年 IT 安全产业中 IT 安全硬件占半壁江山，占比达到 50.50%；软件和服务所占比重不断上升，占比已达到 10.7%[13]。我国自主开发的网络安全产品已涵盖物理安全、通信安全、数据安全、应用安全、安全管理平台以及新技术新应用安全等领域，形成了从安全芯片、网络与边界安全产品、数据安全产品、应用安全产品到安全服务较为完善的网络安全产业结构。2013—2016 年我国 IT 安全产业结构如图 6-5 所示。

---

12 数据为新华三销售收入数据。

13 数据来源：中国电子信息产业发展研究院，《2016—2017 年中国网络安全发展蓝皮书》。

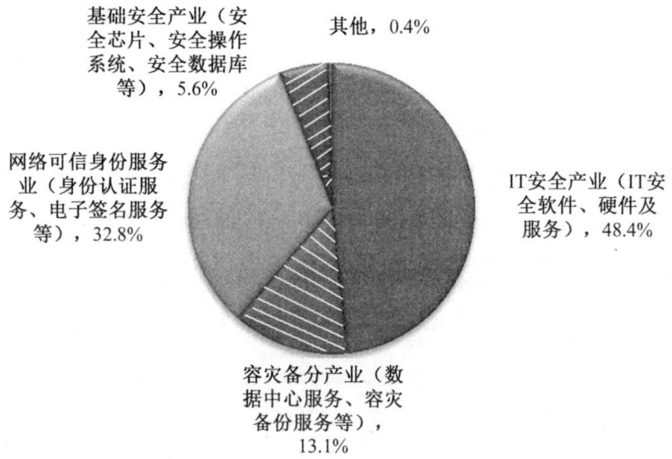

图 6-4　2016 年我国网络安全产业结构

（数据来源：赛迪智库）

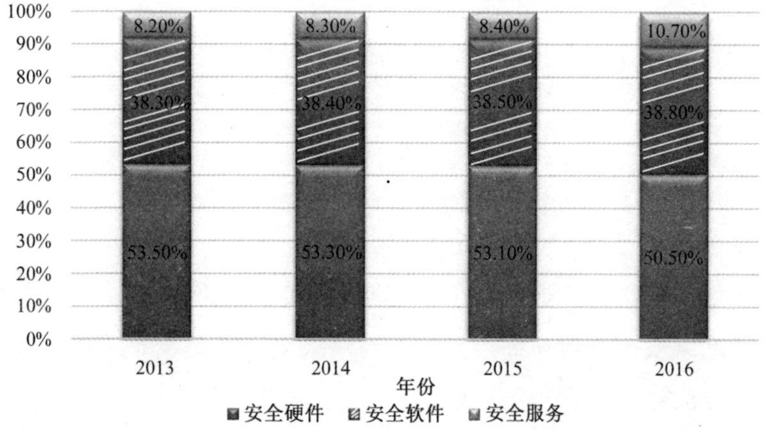

图 6-5　2013—2016 年我国 IT 安全产业结构

（数据来源：赛迪智库）

3．网络安全产业链基本形成

目前，网络安全产业链上下游主要包含高校/科研院所/国家实验室等学术研究机构、基础软硬件生产厂商、平台数据集成厂商及终端用户（见图 6-6）。学术研究机构属于产业链上游，主要负责网络安全相关基础理论研究及核心技术攻关；基础软硬件生产厂商属于中游，负责科研成果产业化，积极研制网络安全基础软硬件；平台数据集成厂商和终端用户属于产业链下游，集成商在基础软硬件的基础上进行二次开发，成为功能各异的终端产品，并提供给终端用户使用。

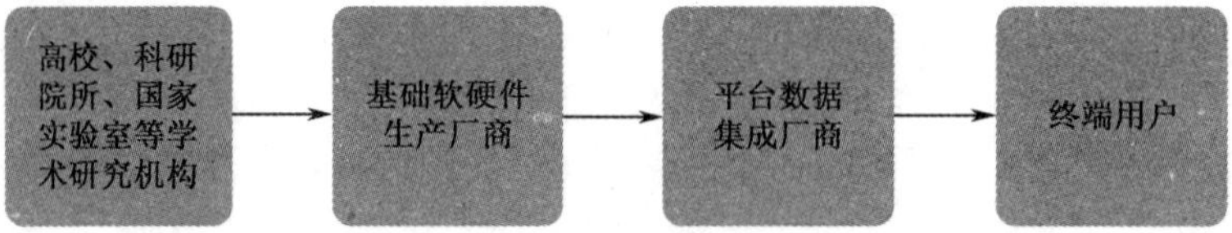

图 6-6　网络安全产业链分布

（数据来源：赛迪智库）

4．网络安全产业园区日益增多

为抓紧网络安全产业发展机遇，不少省市依托相关产业基础和要素，不断加快网络安全产业布局，打造各类网络安全产业专业园区。自 2000 年起，科技部陆续授予三个国家信息安全成果产业化基地，分别为国家信息安全成果产业化（四川）基地、国家信息安全成果产业化（湖北）基地、国家信息安全成果产业化（上海）基地。2016 年 4 月，四川与中国电子科技集团公司开展深度合作，投资 130 亿元建设“中国电科（成都）网络信息安全产业园”，依托“中国电科”子集团——“中国网安”在密码、信息安全、物理安全领域的优势，致力于打造国家级军民融合网络信息安全产业园，重点发展信息安全、安全信息系统、安全运营、安全系统集成、安全服务与测评等业务。2016 年 9 月，烟台市政府也与“中国电子”开展合作，以电子六所研发及产业化中心、企业孵化器等为核心,共建“烟台国家信息安全产业园”，目前首批入园项目已确定，包括金融多功能存取款一体机（CRS）项目、微波通信电子项目、信息安全防护产品研发及产业化项目。湖北省武汉市推动国家网络安全人才与创新基地建设，包括网络安全创新基地、公共孵化中心、网络安全研究院、网络安全人才培训中心、网络安全学院、人才社区、超算中心及共享中心等八大功能区，其中国家网络安全学院于 2017 年 8 月动工开建。2017 年 10 月，工信部与北京市签署《关于建设国家网络安全产业园区战略合作协议》，准备共建国家网络安全产业园区，推动网络安全产业高端化、自主化发展。此外，杭州、合肥、长沙等城市也积极布局网络安全产业，建立网络安全产业园，进一步加大产业发展扶持。

6.3.3 网络安全标准化建设加快

网络安全标准是网络安全保障体系建设的重要技术支撑，是维护国家利益和保障国家安全的重要工具。在经历了萌芽、起步和快速发展三个阶段之后，我国网络安全标准体系已基本构建形成，推动标准化工作常态化、持续化发展。2002 年 4 月，全国信息安全标准化技术委员会（以下简称“信安标委”，委员会编号为〔TC260〕）正式成立，全面规划和管理我国网络安全国家标准。截至 2017 年 11 月，全国信息安全标准化技术委员会发布的信息安全国家标准数量达到 207 个，整体呈现上升趋势。目前，我国实质性参与了多项国际标准化工作，在密码算法、实体鉴别、信息安全事件管理、云计算与虚拟化、大数据安全、可信供应链、生物特征识别等方面均形成了相应成果。1995—2016 年全国信息安全标准化技术委员会历年发布的国标数量如图 6-7 所示。

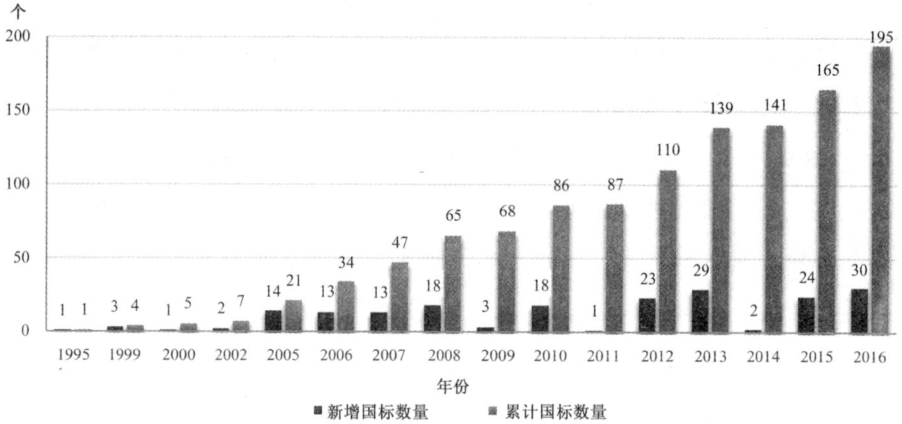

图 6-7　1995—2016 年全国信息安全标准化技术委员会历年发布的国标数量

（数据来源：赛迪智库）

专栏 23　信安标委机构与功能设置

全国信息安全标准化技术委员会负责组织开展国内信息安全有关的标准化技术工作，目前设立八个工作组，主要工作范围包括安全技术、安全机制、安全服务、安全管理、安全评估等领域的标准化技术工作[14]（见图 6-8）。机构设置如下：

信息安全标准体系与协调工作组负责研究信息安全标准体系；跟踪国际信息安全标准发展动态；研究、分析国内信息安全标准的应用需求；研究并提出新工作项目及工作建议。**涉密信息系统安全保密标准工作组**负责研究提出涉密信息系统安全保密标准体系；制定和修订涉密信息系统安全保密标准，以保证我国涉密信息系统的安全。**密码技术工作组**负责密码算法、密码模块，密钥管理标准的研究与制定。

14 信息来源：全国信息安全标准化技术委员会官网（www.tc260.org.cn/）。

鉴别与授权工作组负责国内外 PKI/PMI 标准的分析、研究和制定。**信息安全评估工作组**负责调研国内外测评标准现状与发展趋势；研究提出测评标准项目和制订计划。**通信安全标准工作组**负责调研通信安全标准现状与发展趋势，研究提出通信安全标准体系，制定和修订通信安全标准。**信息安全管理工作组**负责信息安全管理标准体系的研究，信息安全管理标准的制定工作。**大数据安全标准特别工作组**负责大数据和云计算相关的安全标准化研制工作。

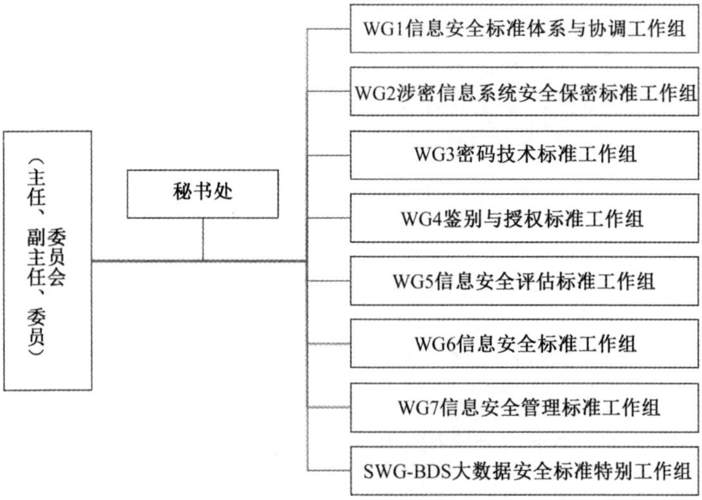

图 6-8　信安标委机构

1．《信息安全标准体系》绘制标准化蓝图

信安标委联合公安部、安全部、国家保密局、国家密码管理委员会、认监委、总参等相关部门，对现有信息安全标准进行了归类和整理，在跟踪分析了国际信息安全标准的发展动态和国内信息安全标准需求的基础上，提出系统、全面、分布合理的信息安全标准体系，经过多次修改，目前已形成《信息安全标准体系》2.0 版，提出了标准体系框架和标准体系表，将信息安全标准划分为七大类：基础标准、技术与机制标准、管理标准、

测评标准、密码标准、保密标准和通信安全标准（见图 6-9）。信息安全标准体系是由信息安全领域内具有内在联系的标准组成的科学有机整体，成为现有、应有和预计制定信息安全标准的蓝图。

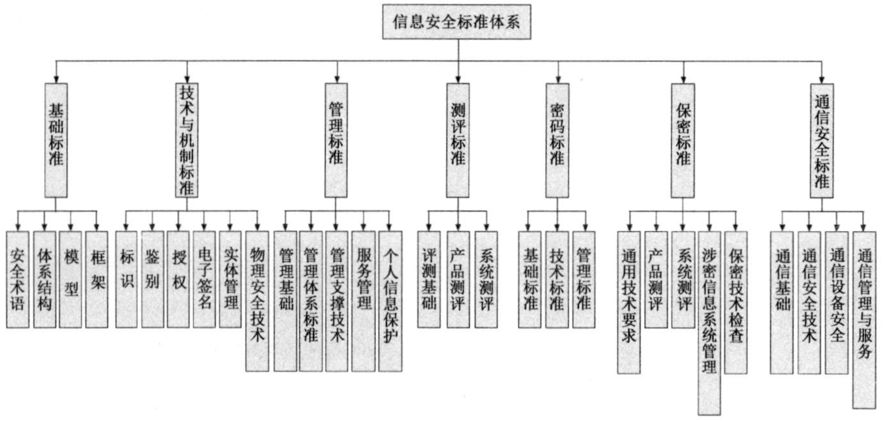

图 6-9　信息安全标准体系总体框架

（数据来源：许玉娜，罗锋盈. 我国信息安全标准体系解析，保密科学技术，2014(1)：17-20）

2. 网络安全标准化进程持续推进

在《信息安全标准体系》的框架基础上，近年来，我国持续关注网络安全领域的标准建设工作，深入推进网络安全标准化进程。2016 年 8 月，中央网信办、国家质检总局、国家标准委联合印发《关于加强国家网络安全标准化工作的若干意见》，对推动我国网络安全标准化进程做出部署，要求建立重大工程、重大科技项目标准信息共享机制，推动国家网络安全相关重大工程或科研项目成果转化为国家标准；要坚持“急用先行”原则，推进急需重点标准制定，开展关键信息基础设施保护、网络安全审查、大数据安全、个人信息保护、新一代通信网络安全、互联网电视终端产品安全、网络安全信息共享等领域的标准研究和制定工作。

6.3.4 网络安全人才培养体系逐步健全

人才是网络安全的第一资源，网络空间的竞争归根结底是人才竞争，建设网络强国需要一支优秀的人才队伍。据统计，中国重要行业信息系统和信息基础设施对于各类网络安全人才的需求量超过 70 万人，预计到 2020 年需求将会达到约 140 万人。然而，截至 2016 年，高校培养的网络安全相关专业人才不到 9 万人，网络安全相关专业每年本硕博毕业生只有约 1 万人，人才缺口依旧巨大[15]。按照习近平总书记提出“下大功夫、下大本钱，请优秀的老师，编优秀的教材，招优秀的学生，建一流的网络空间安全学院”的指示要求，中国正加快网络安全人才体系建设工作。

1. 网络安全人才政策逐步完善

自 2005 年以来，中国政府先后发布《关于进一步加强信息安全学科、专业建设和人才培养工作的意见》《国务院关于大力推进信息化发展和切实保障信息安全的若干意见》等一系列涉及网络安全人才培养的文件，并于 2007 年组建教育部高等学校信息安全类专业教学指导委员会，不断加强信息安全学科建设，培养一支高素质的信息安全人才队伍。2016 年 6 月，中央网信办、国家发改委、教育部、科技部、工信部、人社部六部门联合发布《关于加强网络安全学科建设和人才培养的意见》，提出加快网络安全学科专业和院系建设、创新网络安全人才培养机制、加强网络安全教材建设、强化网络安全师资队伍建设、推动高等院校与行业企业合作育人协同

---

15 数据来源：中国电子信息产业发展研究院，《2016—2017 年中国网络安全发展蓝皮书》。

创新、加强网络安全从业人员在职培训、加强全民网络安全意识与技能培养、完善网络安全人才培养配套措施八条意见，为实施网络强国战略、维护国家网络安全提供强大的人才保障。

2．教育体系逐步成熟

我国信息安全学科经过十余年发展，逐渐形成了独立的基础理论体系、技术体系和应用体系，以及密码学及应用、系统安全、网络安全、应用系统安全等若干二级学科研究方向。2015 年 6 月，国务院学位委员会、教育部联合印发《关于增设网络空间安全一级学科的通知》，在“工学”门类下增设“网络空间安全”一级学科，授予“工学”学位。2016 年，国内网络安全、信息对抗、保密管理 3 个专业共布点 121 个，电子信息类、计算机类等与网络人才培养相关专业类共布点 4 800 余个[16]。目前，网络空间安全一级学科博士学位授权点的高校共 29 所（见表 6-2）；全国 21 所高校成立网络空间安全学院[17]。同时，2017 年 8 月，中央网信办、教育部联合发布《一流网络安全学院建设示范项目管理办法》，在 2017—2027 年实施一流网络安全学院建设示范项目，探索网络安全人才培养新思路、新体制、新机制，形成 4～6 所国内公认、国际上具有影响力和知名度的网络安全学院。目前，已评选公布了首批 7 个一流网络安全学院建设示范项目高校，分别为西安电子科技大学、东南大学、武汉大学、北京航空航天大学、四川大学、中国科学技术大学、战略支援部队信息工程大学[18]。

---

16 数据来源：中国青年报，《中国首个网络安全专项基金成立》，2016 年 2 月 3 日。

17 数据来源：新华社，《为网上家园筑起安全堤坝——网络安全事业五年发展综述》，2017 年 9 月 15 日

18 数据来源：新华网，《首批一流网络安全学院建设示范项目高校名单公布》，2017 年 9 月 16 日。

表 6-2 新增列网络空间安全一级学科博士学位授权点名单

序 号	高校名称	备注	序 号	高校名称	备 注
1	清华大学	新增列	16	华中科技大学	新增列
2	北京交通大学	新增列	17	中山大学	新增列
3	北京航空航天大学	新增列	18	华南理工大学	新增列
4	北京理工大学	新增列	19	四川大学	新增列
5	北京邮电大学	新增列	20	电子科技大学	新增列
6	哈尔滨工业大学	新增列	21	西安交通大学	新增列
7	上海交通大学	新增列	22	西北工业大学	新增列
8	南京大学	新增列	23	西安电子科技大学	新增列
9	东南大学	新增列	24	中国科学院大学	新增列
10	南京航空航天大学	新增列	25	国防科学技术大学	新增列
11	南京理工大学	新增列	26	解放军信息工程大学	新增列
12	浙江大学	新增列	27	解放军理工大学	新增列
13	中国科学技术大学	新增列	28	解放军电子工程学院	对应调整
14	山东大学	新增列	29	空军工程大学	对应调整
15	武汉大学	新增列	—	—	—

数据来源：教育部

3. 网络安全竞赛活动逐步丰富

网络安全竞赛活动是网络安全人才发现、培养和选拔的重要手段，也是完善网络安全教育培训体系中的关键部分，通过各类网络安全竞赛活动，切实提升了网络安全人才的技术水平，发现和涌现出一大批网络安全人才。一是政府部门积极组织信息安全和网络对抗竞赛活动。教育部高等学校信息安全类专业教学指导委员会自2008年起开展全国大学生信息安全竞赛，目前已成功举办了10届，以赛代练、以练促学，培养学生的网络安全创新意识与团队合作精神，普及网络安全知识。从2014年起，连续举办国际网络安全技术对抗联赛（XCTF）。该比赛由网络空间安全人才基金和国家创新与发展战略研究会联合主办，南京赛宁信息技术有限公司总体承办，面向高校及科研院所学生、企业技术人员、网络安全技术爱好者等群体。上海自 2009 年起主办 ISG 信息安全技能竞赛。湖南举办“湖湘杯”2016 网络安全技能大赛。二是高校积极举办相关对抗技术竞赛。如北京理工大学从 2004 年开始举办信息安全与对抗技术竞赛（ISCC）、西安电子科技大学自 2009 年开始主办全国网络安全大赛（XDCTF）等。三是 BAT、奇虎 360 等互联网或网络安全企业大量举办信息安全竞赛。腾讯安全于 2017 年 2 月发起腾讯信息安全争霸赛（TCTF），分为国际赛（0CTF）和新人邀请赛（RisingStar CTF）两大板块，题目涉及破解、逆向、密码学、Web 安全、编程、移动安全等领域。阿里巴巴主办安全技术竞赛 ALICTF，覆盖无线端安全、Linux 内核安全、VPC 网络安全、数据保护安全、Web 安全、逆向工程、安全数据分析、代码安全等众多领域。

4. 网络安全专项基金成立

2016 年 2 月，国内首个网络安全领域的专项基金——中国互联网发展基金会网络安全专项基金捐资仪式在京举行，网络安全专项基金启动资产达 3 亿元，主要用于奖励网络安全优秀人才、优秀教师、优秀标准、优秀教材，资助网络安全专业优秀学生的学习和生活，支持网络安全人才培养基地等重要工作。目前，已举行两届“网络安全杰出人才奖”、“网络安全人才奖”、“网络安全优秀教师奖”、“网络安全奖学金”等评选活动，共评选出网络安全杰出人才 1 名、优秀人才 20 名、优秀教师 18 名，资助网络安全优秀本科生和研究生共 275 名[19]。

19 数据来源：中国互联网发展基金会。

6.4　网络安全防护能力与意识显著提高

网络安全为人民，网络安全靠人民。在网络安全管理和技术的双重保护下，我国着重加强关键信息基础设施防护，并致力于提升全社会网络安全意识，共同维护网络安全，共筑网络安全防线。

6.4.1　关键信息基础设施防护能力持续增强

金融、能源、电力、通信、交通等领域的关键信息基础设施是经济社会运行的神经中枢，是国家的重要战略资源，是网络安全的重中之重。《网络安全法》提出“关键信息基础设施”的概念，并对其范围进行了明确，指出“国家对公共通信和信息服务、能源、交通、水利、金融、公共服务、电子政务等重要行业和领域，以及其他一旦遭到破坏、丧失功能或者数据泄露，可能严重危害国家安全、国计民生、公共利益的关键信息基础设施，实行重点保护”。2017 年 7 月，国家网信办就《关键信息基础设施安全保护条例》公开征求意见，进一步明确了关键信息基础设施保护范围。同时，网络安全态势感知和应急处置能力得到明显提升。2017 年 6 月，中央网信办修订《国家网络安全事件应急预案》，围绕提升网络安全态势感知、事件分析、追踪溯源及遭受攻击后的快速恢复能力，进一步健全完善网络安全监测预警、应急处置、预防等机制；建立完善跨行业、跨部门的威胁情报共享机制，鼓励重点行业、有条件的省（直辖市、自治区）先行建立信息共享机制，探索建设技术支撑平台，并与国家级的平台实现互联互通；加强网络安全事件应急演练，制订完善相关应急预案，提高应对网络安全事件的能力。国家网络安全重要防护对象的演变见表 6-3。

表 6-3 国家网络安全重要防护对象的演变

时 间	文件名称	表 述	涵盖范围
2004 年 9 月	《关于信息安全等级保护工作的实施意见》	国家重点保护涉及国家安全、经济命脉、社会稳定的基础信息网络和重要信息系统	国家事务处理信息系统（党政机关办公系统）；财政、金融、税务、海关、审计、工商、社会保障、能源、交通运输、国防工业等关系到国计民生的信息系统；教育、国家科研等单位的信息系统；公用通信、广播电视传输等基础信息网络中的信息系统；网络管理中心、重要网站中的重要信息系统和其他领域的重要信息系统
2012 年 7 月	《国务院关于大力推进信息化发展和切实保障信息安全的若干意见》	重点领域信息安全	（1）重要信息系统和基础信息网络安全：包含能源、交通、金融等领域涉及国计民生的重要信息系统和电信网、广播电视网、互联网等基础信息网络；无线电；互联网网站、地址、域名和接入服务单位。 （2）政府和涉密信息系统。 （3）工业控制系统。 （4）信息资源和个人信息
2013 年	《关于开展国家重要信息系统调查工作的函》	国家重要信息系统	信息安全等级保护四级信息系统、以及第三级信息系统中跨省全国联网的大系统，本行业、本部门重要业务起到关键支撑性作用或面向公众提供大范围服务（不包括政府网站）的信息系统，以及涉及国家安全、经济命脉、社会稳定和公共利益的极端重要系统
2016 年 11 月	《网络安全法》	关键信息基础设施	公共通信和信息服务、能源、交通、水利、金融、公共服务、电子政务等重要行业和领域，以及其他一旦遭到破坏、丧失功能或者数据泄露，可能严重危害国家安全、国计民生、公共利益的关键信息基础设施

续表

时 间	文件名称	表 述	涵盖范围
2017年7月	《关键信息基础设施安全保护条例（征求意见稿）》	关键信息基础设施	下列单位运行、管理的网络设施和信息系统，一旦遭到破坏、丧失功能或者数据泄露，可能严重危害国家安全、国计民生、公共利益： （1）政府机关和能源、金融、交通、水利、卫生医疗、教育、社保、环境保护、公用事业等行业领域的单位； （2）电信网、广播电视网、互联网等信息网络，以及提供云计算、大数据和其他大型公共信息网络服务的单位； （3）国防科工、大型装备、化工、食品药品等行业领域科研生产单位； （4）广播电台、电视台、通讯社等新闻单位； （5）其他重点单位

6.4.2 网络安全意识明显提升

为更好地帮助公众增强网络安全意识、提高网络安全防护技能，近年来，中国通过举办国家网络安全宣传周等形式，不断加强网络安全宣传教育，营造出网络安全人人有责、人人参与的良好氛围。自2014年以来，中央网信办会同中宣部、中编办、教育部、工信部、公安部、科技部、人民银行、新闻出版广电总局、中国科协、全国总工会、共青团中央等10多个部委，先后联合举办了四届国家网络安全宣传周，围绕共建网络安全、共享网络文明，组织开展了标识推广、知识竞答、专家访谈、短片征集展映等系列主题宣传活动，倡导“网络安全为人民，网络安全靠人民”；直接参与人数过亿，信息覆盖达20亿人次，全民网络安全意识和技能明显提升。调查显示，已有98%的网民认为网络安全“非常重要”[20]，网络安全意识已深入人心。

20 数据来源：新华网，网络调查《增强自我保护，你的网络安全意识如何？》。

第 7 章　积极参与网络空间国际治理

互联网拓展了国家治理新疆域，创造了人类生活新空间，让世界变成了“鸡犬之声相闻”的地球村。但全球网络空间发展不平衡、规则不健全、秩序不合理等问题日益凸显，信息鸿沟、侵犯知识产权、网络犯罪、网络监听、网络攻击、网络恐怖主义等一系列治理难题层出不穷，对国家主权、安全和发展利益提出新挑战。中国国家主席习近平高度重视网络空间国际治理，在多个重要国际场合阐述国际治网理念，特别是提出了推进全球互联网治理体系变革“四项原则”和构建网络空间命运共同体“五点主张”，推动构建和平、安全、开放、合作的网络空间，建立多边、民主、透明的全球互联网治理体系。

7.1　网络空间国际治理中国主张赢得广泛认同

7.1.1　网络空间国际治理中国理念日益丰富

中国积极参与网络空间国际治理，推动世界各国在相互尊重、相互信任的基础上，共谋发展、共筑安全。2014 年 5 月，在上海举行的亚信峰会第一阶段会议中，习近平主席提出安全应该是普遍的，不能牺牲别国安全谋求自身所谓“绝对安全”。2014 年 7 月，习近平主席在巴西国会演讲时指出，中国愿意同世界各国携手努力，尊重网络主权，维护网络安全，共同构建和平、安全、开放、合作的网络空间，建立多边、民主、透明的国际互联网治理体系。这是世界上首次由国家最高领导人明确提出网络主权原则。2015 年 12 月，在第二届世界互联网大会开幕式主旨演讲中，习近平主席提出推动全球互联网治理体系变革“四项原则”和构建网络空间命运共同体“五点主张”，完整系统地阐述了互联网治理的“中国主张”，在世界互联网史上具有里程碑意义。2016 年 11 月，在第三届世界互联网大会开幕式上，习近平主席进一步强调携手构建网络空间命运共同体，表示“中国愿同国际社会一道，坚持以人类共同福祉为根本，坚持网络主权理念，推动全球互联网治理朝着更加公正合理的方向迈进，推动网络空间实现平等尊重、创新发展、开放共享、安全有序的目标”。

专栏 24 “四项原则”和“五点主张”

2015 年 12 月 16 日，国家主席习近平在第二届世界互联网大会开幕式主旨演讲中，在深刻分析当前世界互联网发展阶段特点的基础上，提出互联网发展的“四项原则”和“五点主张”。

“四项原则”：第一，坚持尊重网络主权，尊重各国自主选择网络发展道路、网络管理模式、互联网公共政策和平等参与国际网络空间治理的权利。第二，坚持维护和平安全，网络空间不应成为各国角力的战场，更不能成为违法犯罪的温床，维护网络安全不应有双重标准。第三，坚持促进开放合作，创造更多利益契合点、合作增长点、共赢新亮点，推动彼此在网络空间优势互补、共同发展，让更多国家和人民搭乘信息时代的快车、共享互联网发展成果。第四，坚持构建良好秩序，依法治网、依法办网、依法上网，同时要加强网络伦理、网络文明建设，发挥道德教化引导作用。

“五点主张”：第一，加快全球网络基础设施建设，促进互联互通，让更多发展中国家和人民共享互联网带来的发展机遇。第二，打造网上文化交流共享平台，促进交流互鉴，推动世界优秀文化交流互鉴，推动各国人民情感交流、心灵沟通。第三，推动网络经济创新发展，促进共同繁荣，促进世界范围内投资和贸易发展，推动全球数字经济发展。第四，保障网络安全，促进有序发展，推动制定各方普遍接受的网络空间国际规则，共同维护网络空间和平安全。第五，构建互联网治理体系，促进公平正义，应该坚持多边参与、多方参与，更加平衡地反映大多数国家意愿和利益。

7.1.2 网络空间国际合作战略性文件发布

2017 年 3 月，《网络空间国际合作战略》（以下简称“战略”）发布。战略以和平发展、合作共赢为主题，以构建网络空间命运共同体为目标，全面系统阐释了推动网络空间国际交流合作的中国主张。战略倡导各国切实遵守《联合国宪章》宗旨与原则，确保网络空间的和平与安全；坚持主权平等，不搞网络霸权，不干涉他国内政；各国共同制定网络空间国际规则，建立多边、民主、透明的全球互联网治理体系；推动在网络空间优势互补、共同发展，弥合“数字鸿沟”，确保人人共享互联网发展成果。战略还从九个方面提出了中国推动并参与网络空间国际合作的行动计划，并强调中国在推动建设网络强国战略部署的同时，将秉持以合作共赢为核心的新型国际关系理念，致力于与国际社会携起手来，加强沟通交流，深化互利合作，构建合作新伙伴，同心打造人类命运的共同体，成为指导中国参与网络空间国际交流与合作的纲领性文件，为世界各国网络空间合作提供了参考。

7.2 搭建互联网治理国际平台

中国积极发挥负责任大国作用，搭建全球性、区域性多层次互联网治理平台，全面宣介中国治网主张，推动世界各国特别是发展中国家的政府、企业、民间团体等在互联网国际治理领域积极发声、密切交流、凝聚共识、深化合作，为推进全球互联网治理体系变革、构建网络空间命运共同体发挥积极作用。

7.2.1 创办世界互联网大会

世界互联网大会自2014年创办以来，日益成为中国搭建的最具代表性的互联网国际治理平台，在思想交流、技术展示、经贸合作、凝聚共识等方面取得丰硕成果，搭建了中国与世界互联互通的国际平台、国际互联网共享共治的中国平台。从2014年到2016年，三届大会共吸引4 600多位各国政府官员、专家学者、企业家和国际机构负责人参会[1]，在全球范围内产生巨大影响力。2014年11月，首届“世界互联网大会”在浙江乌镇举办，习近平主席向首届世界互联网大会致贺信，指出互联网发展对国家主权、安全、发展利益提出了新的挑战，迫切需要国际社会认真应对、谋求共治、实现共赢。2015年12月，习近平主席出席第二届世界互联网大会开幕式并作主旨演讲，提出互联网发展治理“四项原则”、“五点主张”。大会秘书处高级别专家咨询委员会（以下简称“高咨委”）正式成立，由马云和前任ICANN总裁兼首席执行官法迪 • 切哈德（Fadi Chehadé）担任联合

1 数据来源：三届互联网大会闭幕式新闻发布会公开发布数据。

主席，集聚全球互联网产业界、学术界、研究机构领军人物为世界互联网发展献计献策。经高咨委讨论并通过，大会组委会发布《乌镇倡议》，体现了“开放、平等、协作、分享”的互联网精神，得到全球广泛关注。大会同期举办的“互联网之光”博览会，吸引了全球近 260 家企业参展，举办了 80 多场各具特色的专场发布会。2016 年 11 月，以“创新驱动造福人类——携手共建网络空间命运共同体”为主题的第三届世界互联网大会召开，国家主席习近平在大会开幕式上发布视频讲话。大会首次举办世界互联网领先科技成果发布会，囊括人工智能、量子计算、网络安全等领域全球最前沿、最具影响力的科技创新成果，带来互联网领域的尖端技术盛宴。大会发布了《2016 年世界互联网发展乌镇报告》，全面总结一年来世界互联网发展与治理经验成就，深入分析互联网发展面临的形势，提出下一步全球互联网发展与治理重点，持续推动“乌镇进程”。

7.2.2 积极构建国际合作双边多边对话体系

中国积极推动建立互联网领域双边多边对话机制，持续推进与世界各国网络空间深度合作，开辟国际交流的新通道，搭建起灵活、多元的网络空间国际治理合作对话体系。

1. 不断加强中美网络空间交流对话

中美作为网络大国，两国在网络空间的交流合作不仅关乎两国的利益，更直接影响到国际网络空间治理体系的变革发展。近年来，中国通过一系列论坛会议，与美政府、互联网企业、智库等构建起立体交流合作机制。自 2007 年起，中美互联网论坛已成功举办八届。2015 年 9 月，习近平主席会见出席第八届中美互联网论坛代表时强调，中美双方理应在相互尊重、相互信任的基础上，就网络问题开展建设性对话，打造中美合作的亮点，让网络空间更好造福两国人民和世界人民，得到中美两国互联网企业家的普遍认同。自 2015 年 9 月起，中美积极举办网络空间国际规则高级别专家组会议，深入探讨网络空间国际规则问题。2017 年 4 月中美元首会晤，建立执法及网络安全对话等 4 个高级别对话机制，并于 2017 年 10 月举行首轮执法及网络安全对话，就打击网络犯罪等议题进行深入探讨，达成广泛共识和务实成果，中美网络空间对话机制不断巩固发展。

2. 持续深化中俄互联网治理合作

2016 年 6 月，习近平主席与俄罗斯总统普京签署中俄元首《关于协作推进信息网络空间发展的联合声明》，指出包括中俄在内的各国在互联网领域都拥有重要的共同利益与合作空间，理应在相互尊重和相互信任的基础上，就保障信息网络空间安全、推进信息网络空间发展的议题，全面开展实质性对话与合作。中俄双方达成了尊重网络主权、加强技术合作、打击网络犯罪、网络安全应急合作等多项共识。这对于中俄两国乃至世界的互联网交流合作具有十分重要的指导意义，成为新型大国关系在国际网络空间合作中的典范。

3. 广泛开展中欧互联网治理合作

近年来，中欧互联网合作逐步深入，持续推动中欧数字经济和网络安全专家工作组、中欧数字合作圆桌会议、中英互联网圆桌会议、中德互联网产业圆桌会议等对话交流机制建立完善，数字经济、“互联网+”、5G 技术、儿童在线保护等成为中欧交流合作新亮点，对促进两国政府和业界的互联网合作发挥了积极作用。

4. 密切与其他发展中国家互联网交流合作

成功举办“中国—东盟信息港论坛”，通过打造中国-东盟信息港引领和带动中国与东盟网络空间的全面交流；举办中国—南非互联网圆桌会议，多层次、多领域深化两国网络空间务实合作；举办“中国—阿拉伯博览会网上丝绸之路论坛”，建设中国—阿拉伯网上丝绸之路宁夏枢纽工程，着力建设造福中阿人民的“网上丝绸之路”，服务“一带一路“建设。

7.3 加强与国际机构的交流合作

中国积极参与互联网国际治理工作，深度参与并影响互联网国际治理进程，在联合国、世界经济论坛等全球性的网络空间治理平台和 ICANN（互联网名称和数字地址分配机构）等互联网技术管理机构的发展变革中起到重要作用。

7.3.1 高度重视与国际机构开展合作

1. 积极投身联合国框架下互联网治理工作

信息社会世界峰会（WSIS）及互联网治理论坛（IGF）是联合国框架下最重要的互联网领域对话平台。WSIS 是联合国在近 20 年时间里开展国际网络空间治理的重要成果，中国政府及民间组织参与 WSIS 各个阶段的议程，积极影响全球互联网治理进程。在 2003 年 12 月在瑞士日内瓦举行的第一阶段峰会上，中国代表团共有 19 人参加，并主要围绕“数字鸿沟”等发展问题，强调政府在管理信息社会、解决数字鸿沟的重要性；2005 年 11 月在突尼斯城举行的第二阶段峰会上，中国代表从促进协调发展、加强国际合作、充分尊重各国社会制度差异性和文化多样性四个方面阐述了中国政府对建设信息社会的观点和主张。2015 年，在 WSIS 成果执行情况 10

年审议会议上，中国政府先后提交了三份意见书。中国政府和社会组织还参加了历届 IGF 论坛。2016 年，WSIS 与 IGF 启动新的十年进程，在当年 G20 杭州峰会上，与会国家一致认为互联网治理应继续遵循 WSIS 成果，强调政府、私营部门、民间社会、技术团体和国际组织等各方应根据各自的角色和责任，充分、积极参与国际互联网治理。

2．积极推动 ICANN 国际化进程

ICANN 负责互联网协议地址的空间分配、协议标识符的指派、通用顶级域名与国家和地区顶级域名系统的管理以及根服务器系统的管理。中国一贯重视推动互联网关键基础资源的合理分配，积极承办 ICANN 会议，大力推动 ICANN 管理机制改革，对美国将 IANA 管理权从行政管辖向司法管辖转变起到积极作用。中国互联网络信息中心（CNNIC）已经成为 ICANN 认证亚太唯一的新通用顶级域（New gTLD）数据托管服务机构，为中国乃至整个亚太地区新顶级域注册服务机构和注册者在数据恢复、语言沟通、纠纷仲裁等方面提供更优质和本土化的服务。2015 年 6 月，中国参与 ICANN 第 53 届会议上发布的“雪人计划”并担任计划首任执行，该计划基于 IPv6 等全新技术框架，旨在打破现有国际互联网 13 个根服务器的数量限制，制定更完善的下一代互联网根服务器运营规则。

7.3.2 积极参与互联网国际组织工作

1．中国专家积极担任国际组织重要职务

中国互联网协会理事长胡启恒院士，于 2004 年接受时任联合国秘书长安南邀请，成为联合国互联网治理工作组（WGIG）小组 40 名成员之一，代表中国互联网社群作出积极贡献，并于 2013 年成为中国首位入选“互联

网名人堂”的杰出贡献人士。中科院计算机网络信息中心首席科学家钱华林，2003 年作为中国专家第一次当选 ICANN 理事，并曾担任亚太地区顶级域联合组织（APTLD）主席，2014 年他成为中国入选“互联网名人堂”第二人。赵厚麟于 2014 年成为国际电信联盟（ITU）150 年历史上首位中国籍秘书长。清华大学教授李星长期参与国际互联网工程任务组（IETF）研究工作，于 2013 年入选新一届互联网体系架构委员会（IAB）。下一代互联网工程中心主任刘东 2015 年担任 IEEE 标准协会董事。此外，有多位中国专家在 ICANN 和 IETF 担任重要职务，为推动互联网国际治理发挥了积极作用。

2. 企业积极推动国际事务合作进程

中国企业发挥自身的技术特长，加入各类互联网领域的国际组织。华为是中国企业参与国际互联网标准化进程的代表，截至 2016 年 12 月，华为加入了 360 多个标准组织/产业联盟/开源社区，担任 300 多个重要职位，在 IEEE-SA 标准委员会、WFA（Wi-Fi 联盟）、OneM2M（物联网协议联盟）、OMA（开放移动联盟）、OASIS（结构化信息标准促进组织）等国际组织中担任董事会成员，获得多个重要职位并主导关键立项。百度是中国首个受邀加入万维网联盟（W3C）的企业，积极参与网络开发平台和无线互联网络等领域标准的制定；截至 2016 年 12 月，奇虎 360、腾讯、阿里巴巴、中国移动等 32 家企业成为 W3C 会员[2]。

2 数据来源：万维网联盟中国办事处官方网站更新数据（http://www.chinaw3c.org/china-members.html）。

第8章　互联网发展造福人民

习近平总书记指出，网信事业要发展，必须贯彻以人民为中心的发展思想。中国互联网始终坚持把为人民服务、增进人民福祉、促进人的全面发展作为互联网发展的出发点和落脚点。加快信息化服务普及，降低互联网应用成本，开展网络扶贫，为老百姓提供用得上、用得起、用得好的信息服务。充分利用互联网普惠、便捷、共享特性，促进资源共享，提升公共服务水平，让亿万人民在共享互联网发展成果上有更多获得感。

8.1　互联网普及促进网民均衡发展

经过 23 年的发展，中国网民数量达 7.51 亿人，居全球之首，互联网普及率达 54.3%。互联网为不同性别、不同年龄、不同学历、不同职业、不同收入、不同健康条件的网民，提供了均等地了解世界、获取信息、实现自身发展的机会。庞大的网民基础，也为中国互联网持续发展提供了坚实基础，成为实现中国互联网创新发展的力量源泉，成为推动互联网发展惠及更多人民群众的不竭动力。

8.1.1　网民规模快速增长

1. 中国网民规模持续稳健增长

中国网民发展可以分为两个时期：一是高速增长期，1994—2008 年，中国网民规模持续高速发展。1994 年中国实现全功能接入互联网之际，只有少量的科技工作者、科研技术人员能够上网；到 2002 年，中国网民数量达 5 910 万人，世界排名第二，仅次于美国；2005 年 6 月，网民规模突破 1 亿人，互联网服务市场的用户规模效应初步形成，互联网产业开始快速发展创新；2008 年 6 月，网民规模达 2.53 亿人，跃居世界第一位。二是增长平台期，从 2009 年开始中国网民规模增速放缓，进入数量增长的平台期。2009 年中国网民规模增速从 2008 年的 41.9%下降到 28.9%，不过从绝对规模上看，由于中国网民基数庞大，依然实现 8 600 万人的增长数量。截至 2017 年 6 月，中国网民规模达 7.51 亿人（见图 8-1），互联网普及率达 54.3%，平均上网时长达到 26.5 小时/周[1]。1997—2017 年 6 月世界与中国互联网普及率如图 8-2 所示。

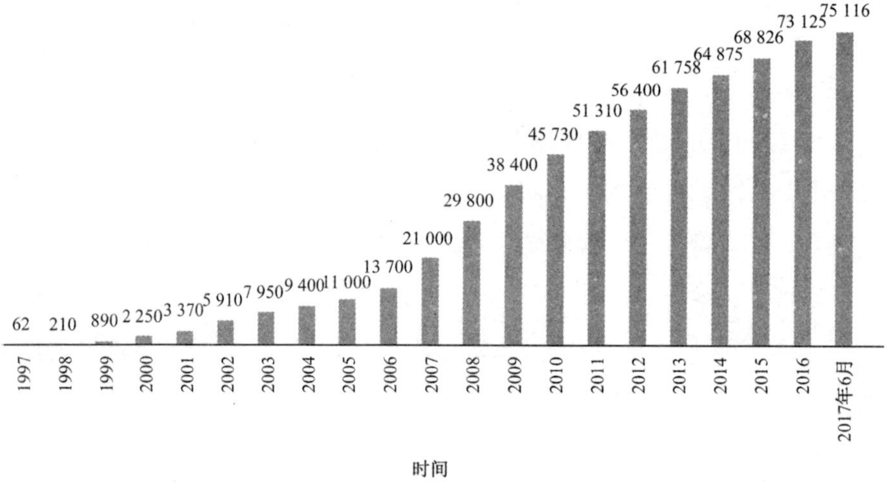

图 8-1　中国网民规模（万人）（1997—2017 年 6 月）

（数据来源：CNNIC）

1 数据来源：CNNIC。

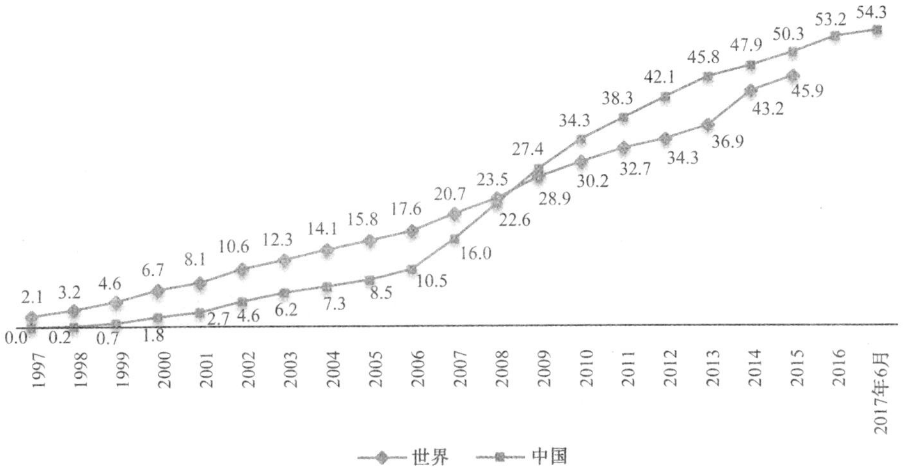

图 8-2　世界与中国互联网普及率（1997—2017 年 6 月）

（数据来源：CNNIC，世界银行）

2．手机网民增长迅速

随着移动互联网的普及，移动应用程序内容质量的提升，网民向移动端转移的趋势明显。从 2009 年开始，移动终端产业繁荣发展、移动通信网络广泛覆盖，特别是 3G 的开通使得手机上网的速率大幅度提升，手机上网成为了潮流，拉动了网民规模实现新一轮平稳增长。根据工业和信息化部数据，全国每百人移动电话拥有量从 1995 年的 0.3 部增长至 2017 年的 98.7 部，一线城市终端设备普及率已趋于饱和，中西部地区的终端普及率快速上升。截至 2017 年 6 月，我国手机网民规模达 7.24 亿人，网民使用手机上网的比例提升至 96.3%[2]。1997—2016 年手机网民增长率与整体网民增长率如图 8-3 所示。

2 数据来源：CNNIC。

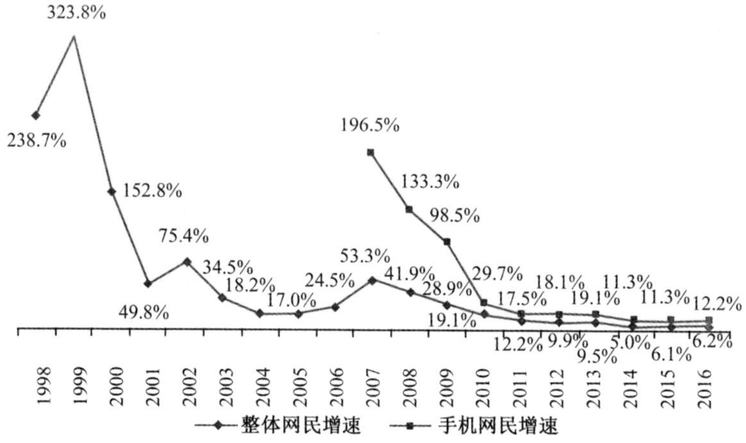

图 8-3 手机网民增长率与整体网民增长率（1997—2016 年）

（数据来源：CNNIC）

8.1.2 互联网加速普及

1. 互联网普及率的省际、城乡差异逐渐缩小

从 1997 年至 2016 年，中国各省区市互联网普及水平提升显著，通过变异系数[3]分析可以看到，互联网普及率的省际差异从 3.37 降低至 0.19（见图 8-4），区域间互联网普及差异呈现稳定下降趋势。中国高度重视农村互联网普及工作，针对当前农村地区尤其是老、少、边、山、穷地区互联网发展不平衡的现象，研究部署一系列改革推进措施。2004 年的“村村通工程”、2014 年的“宽带乡村”工程、2015 年的“改革创新电信普遍服务补偿机制”，均取得卓越成效，实现了我国农村信息基础设施的跨越式发展，城乡“数字鸿沟”问题得到极大改善。根据 2017 年 10 月召开的工业通信

3 变异系数，是一组数据的标准差与其平均值的比，反映该组数据变异程度的大小。省互联网普及率的变异系数越大，说明互联网普及率的地区差异越大。

业发展情况发布会上公布的数据，截至第三季度末，电信普遍服务两批试点总投资超过 400 亿元，共部署 13 万个行政村通光纤，其中有 4.3 万个是建档立卡的贫困村，行政村通宽带的比例已经超过了 96%，贫困村宽带的覆盖率已经达到了 86%。1997 年、2005 年和 2016 年全国各省（自治区、直辖市）的互联网普及率如图 8-5 所示，中国网民分城乡的互联网普及率变动情况如图 8-6 所示。

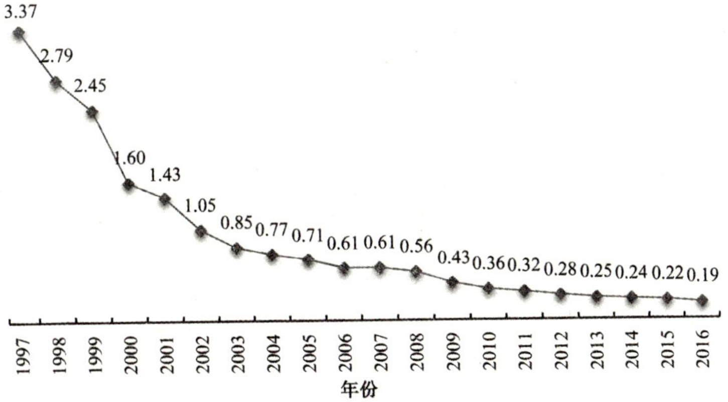

图 8-4　互联网普及率的省间差异（变异系数）变动情况（1997—2016 年）

（数据来源：CNNIC）

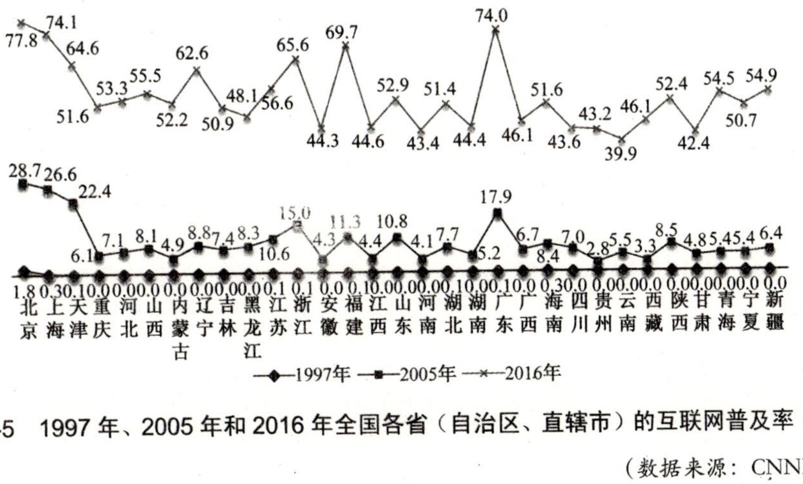

图 8-5　1997 年、2005 年和 2016 年全国各省（自治区、直辖市）的互联网普及率（%）

（数据来源：CNNIC）

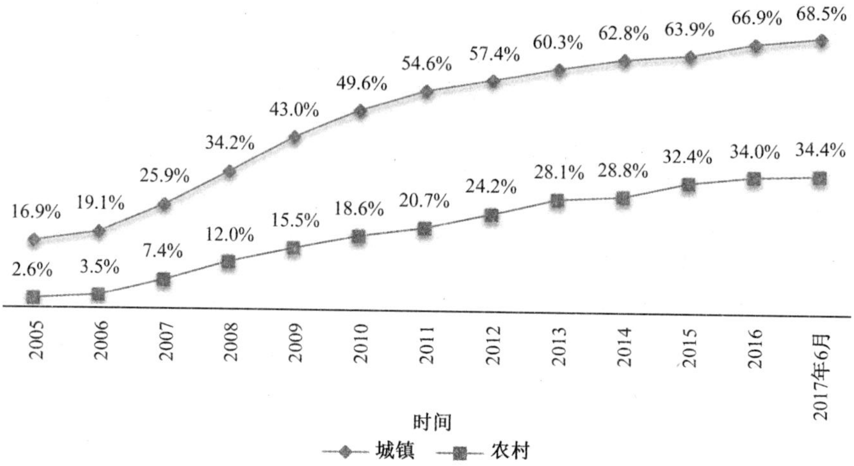

图 8-6　中国网民分城乡的互联网普及率变动情况[4]（2005—2017 年 6 月）

（数据来源：CNNIC）

2．互联网助力精准脱贫

统筹实施“网络覆盖工程、农村电商工程、网络扶智工程、信息服务工程、网络公益工程”五大工程，致力于推进精准扶贫、精准脱贫。充分发挥市场力量，农村电子商务发展成果显著，电子商务进农村综合示范县达到 496 个，带动农民增收超过 100 亿元[5]。

8.1.3　信息无障碍建设取得阶段性进展

2008 年 4 月《残疾人保障法》修订通过，首次将信息无障碍建设纳入

4 2017 年 6 月全国分城乡人口数为估计值，为 2016 年末分城乡人口数+（2016 年末分城乡人口数-2015 年末分城乡人口数）/2。

5 2017 年 1—9 月全国农村电子商务运行情况，http://scjss.mofcom.gov.cn/article/cx/201710/20171002660911.shtml.

法律条款，明确“国家和社会应当采取措施，逐步完善无障碍设施，推进信息交流无障碍，为残疾人平等参与社会生活创造无障碍环境”；同年 7 月，工业和信息化部《信息无障碍 身体机能差异人群网站设计无障碍技术要求》实施，这是中国颁布的第一个信息无障碍的行业技术标准。此后，又相继出台《关于加快推进残疾人社会保障体系和服务体系建设的指导意见》《无障碍环境建设条例》《关于加强网站无障碍服务能力建设的指导意见》等一系列政策法规，大力推动各级政府进行信息交流无障碍建设。2005 年以来，中国残疾人联合会、中国残疾人福利基金会、中国互联网协会等单位发起组织了“十万盲人学电脑公益活动”，并组织国内近百家大型媒体网站成立了中国信息无障碍联盟，在北京奥运会、上海世博会、广州亚运会等重大活动中，组织开展了“网站信息无障碍行动”等活动。企业积极创新信息无障碍技术和产品应用，2013 年腾讯、阿里巴巴、百度等多家企业联合成立“中国信息无障碍产品联盟”，大力发展信息无障碍技术，不断推出受残疾人欢迎的各种辅助软件和新型产品。2017 年中国政务信息无障碍“普及行动”开展，计划在 2018 年年底，全面完成中国政务信息无障碍环境建设工作，使所有人都能平等获取政务信息和服务。

8.2 互联网应用推进普惠民生

8.2.1 互联网医疗推动健康服务高效化

1. 医疗信息化建设取得积极进展

疾病信息警示机制、公共卫生信息系统先后建立并不断完善。2003 年

"非典"疫情爆发后建立了依托互联网的中国疾病预防控制信息系统（CDC），覆盖了传染病监测、突发公共卫生事件报告等主要疾病预防控制业务领域，以及98%的县级以上医疗卫生机构和91%的乡镇卫生院，实现了突发性公共卫生事件、应急信息的准确监测与及时报告，为防控决策、快速响应提供了重要信息支撑。2015—2016 年中国医院信息化状况调研报告显示[6]，医院管理信息和临床医疗管理信息领域的信息化实施程度都已处于较高水平。其中，住院药房管理系统、药库管理系统、门急诊挂号系统等医院管理信息系统（HIS）实施比例为 70%~75%；住院护士、医生工作站信息系统、电子病历系统等临床医疗管理信息系统（CIS）实施比例也均超过 70%。互联网推动医疗健康服务精准化、个性化，专业医疗机构与互联网企业合作，推动在线诊疗、线上健康指导和优质医疗资源共享，各类便携式医疗智能终端、可穿戴设备让移动医疗更加贴近百姓生活。

2．人工智能技术初步应用于医疗领域，大数据提升医疗效率

近年来，互联网企业加快人工智能技术在医疗领域的布局，AI 产品逐步应用于医疗临床诊断和医生辅助。2016—2017 年，百度、阿里巴巴、腾讯等互联网企业分别在医疗领域布局 AI 产品百度医疗大脑、ET 医疗大脑、腾讯觅影。通过收集、汇总、分类、整理患者症状描述，AI 系统获得海量医疗数据，利用人工智能技术分析模拟医生问诊流程，辅助医生完成问诊，同时在医学影像、新药开放、健康管理等领域承担医生助手角色。与此同时，传统医疗机构也积极尝试与互联网平台合作，利用大数据技术提升医疗诊断效率。2017 年，中南大学湘雅二医院、丁香园和大拿科技在皮肤病人工智能辅助诊断达成合作，通过临床数据采集和挖掘既往历史数据，研发出红斑狼疮人工智能辅助诊断模型，目前疾病识别准确率超过 85%[7]。

---

6 调研涉及 342 家三级医院，194 家三级以下医院。

7 中国首个皮肤病人工智能辅助诊断系统发布 "拍图识病"成为现实，http://www.sohu.com/a/141803141_

3. 线上线下融合发展，远程医疗模式蓬勃发展

2014年，中国第一家互联网医院在广东省成立，互联网重塑医疗体系初步开始。2015年底乌镇互联网医院成立后，互联网医院开始迎来建设高峰期，浙大一院互联网医院、广东云医院等超过30家互联网医院先后成立。互联网在线诊疗平台从线上逐步走向线下，增加医疗服务可及性和互动性。2015年开始，春雨医生、丁香园等线上诊疗平台先后成立线下诊所，2017年腾讯“企鹅医生”诊所、杭州“共享医院”先后落地。借助互联网医院和移动医疗平台，远程会诊、远程诊断、远程监护等医疗服务在全国各医疗机构逐步开展。国家卫计委数据显示，截至2016年年底，全国共计6 800余家医疗机构开展远程医疗服务，覆盖了1 330个县，打通整合了跨医院、跨区域间的医疗资源，有效提升医生、医疗设备等资源的使用效率。针对挂号难、候诊时间长、诊疗时间短、支付环节多等困扰患者的医疗痛点，互联网医疗服务逐步向求医问诊、预约挂号、购药环节渗透，以满足用户需求。CNNIC数据显示，网上医疗信息查询、咨询问诊、预约挂号服务用户使用率已分别占互联网医疗用户的40.6%、39.1%和24.1%，有效提升了就医便捷度、服务满意度。

8.2.2 互联网教育助力教育资源均等化

1. “三通两平台”推进教育信息化快速发展

2010年，国务院发布了《国家中长期教育改革和发展规划纲要（2010—

114731

2020 年)》，将发展教育信息化提升到国家战略高度。通过“三通两平台”（宽带网络校校通、优质资源班班通、网络学习空间人人通，以及教育资源公共服务平台、教育管理体服务平台）建设，中国教育信息化取得显著成效。教育部数据显示，2012—2016 年，全国中小学互联网接入率从 25%上升到 87%，多媒体教室比例从不到 40%提高到 80%，每百名中小学生拥有计算机数从 8 台增长到 12 台。上海、江苏等部分东部发达省市已启动无线教育城域网、校园网和“未来教室”建设试点，学校网络教学环境大幅改善。“教育资源公共服务平台”于 2012 年末上线运行，截至 2017 年 8 月，平台已开通教师空间共计 1 136 万个、学生空间 515 万个、家长空间 448 万个，实现与 22 个省级平台和 23 个市/县级平台的互联互通[8]。2016 年，教育部数据中心已建成较为完整的网络、存储、计算、安全和运维架构。32 个省级教育数据中心陆续完成基础设施建设，实现部省两级数据中心网络连接[9]。

2. 大型开放式网络课程蓬勃发展

作为在线教育较早的服务模式之一，世界范围内大规模在线开放课程（慕课，MOOC）在 2013 年开始在中国发展，此后高校主导和企业主导的慕课在线教育模式蓬勃发展。短短 4 年时间，高等教育领域已建成爱课程网“中国大学 MOOC”、清华大学“学堂在线”、上海交通大学“好大学在线”、北京大学“华文慕课”等一批具有一定影响力的在线课程服务平台，上线的慕课超过 1 400 门、为高校定制课程（SPOC）5 600 多门次，选课超过 3 000 万人次，获得在线课程学分的在校生超过 200 万人次。2016 年我国高水平大学又有 40 余门课程登陆国际著名平台，目前已有 170 余门中

8 《2017 年 8 月教育信息化工作月报》，教育部。

9 《2016 年教育信息化工作进展情况》，教育部。

国优质在线开放课程走向世界，992门精品视频公开课和2 886门精品资源共享课在网上免费开放[10]。

3. 在线教育应用产品日趋丰富

直播、录播、微课等服务形式创新不断，在线辅导、题库、作业答疑等各类应用层出不穷，形成了多元化、定位清晰的互联网教育产品，进一步扩大学生参与规模，促进师生互动、教学相长。大数据、人工智能等技术深度影响在线教育，沪江网的“Uni 智能学习系统”、学霸君的“高考机器人”、英语流利说的“AI 英语老师”等产品创新不断，实现在多个教育场景的应用。CNNIC 数据显示，截至 2017 年 6 月，中国在线教育用户规模达 1.44 亿人，半年增长率为 4.8%，在线教育用户使用率为 19.2%。

8.2.3 互联网公益带动传统公益创新发展

从 2007 年开始，支付宝、财付通等第三方网络支付平台，已经开始进入公益慈善领域，改变了慈善机构只能依赖银行汇款和募捐的传统模式，降低了筹款门槛，便利了民众捐赠。2013 年开始，凭借微信、微博等网络社交平台，公益慈善传播的速度得到极大提高，传播成本极大降低，互联网在善款支付、慈善理念传播等方面发挥积极作用，快速推动了中国公益慈善事业的发展。2015 年，腾讯公益创新上线“益行家”线上捐步项目，用户将被记录的运动步数捐出后，可换取鸡蛋、牛奶、饮用水等物资进行捐助，截至 2017 年 10 月，4.7 亿名网友用步数兑换成 5.8 亿元公益基金[11]。

---

10 教育部《2016 年教育信息化工作进展情况》。

11 数据来源：人民日报，《搭上“互联网+”快车，随手公益更接地气》，作者：钱一彬、赵丹彤、邝西曦，2017 年 11 月 2 日。

2016 年 8 月，支付宝“蚂蚁森林”上线，收取用户采用公共交通出行、网上缴费等绿色低碳行为的“绿色能量”，由蚂蚁金服在内蒙古阿拉善等地区代种梭梭、沙柳等防沙绿植，截至 2017 年 8 月底，用户超过 2.3 亿人，累计减排 122 万吨，种植树苗 1 025 万棵，总面积超过 16 万亩[12]。2017 年 8 月，四川九寨沟县地震发生后，中国社会福利基金会等多家机构通过微公益平台发起紧急救援项目，共募集善款超过 300 万元[13]。2016 年，腾讯公益平台、蚂蚁金服公益平台、淘宝公益平台 3 家平台全年筹款达 12.89 亿元，比 2015 年增加 37.79%[14]；从网络捐赠的领域来看，医疗救助和教育助学类最受关注，其次为减灾救灾和环境保护。据不完全统计，截至 2017 年 10 月，通过网络实施的捐赠超过了 10 亿人次，捐赠金额超过 20 亿元，捐赠主体由“80 后”、“90 后”构成；一些慈善组织来自互联网的捐赠占到总收入的 80%以上[15]。

专栏 25 唯爱工坊

“唯爱工坊”是唯品会推出的特色电商公益平台，通过为公益商品提供免费的、成熟的电商平台支持，加速公益商品的市场化，也为 3 亿名会员打造“品质生活，从爱开始”的生活体验。

唯品会前期投入经费组织生产，并免费提供设计、质检、包装、运营、推广、物流等全方位的支持，并在“唯爱工坊”进行零利润销售，所得全额用于帮助生产公益商品的贫困妈妈等社会弱势群体，让

---

12 数据来源：阿里 95 公益周。

13 数据来源：人民日报，《动动手指做公益越来越火》，作者：钱一彬、赵丹彤、邝西曦，2017 年 11 月 2 日，http://paper.people.com.cn/rmrb/html/2017-11/02/nw.D110000renmrb_20171102_1-14.htm.

14 数据来源：《2016 年度中国慈善捐助报告》。

15 央广网，《民政部：网络募捐成为慈善新模式 募捐主体多为 80、90 后》，记者刘会民，2017 年 10 月 15 日。

她们能够获得一份有尊严且可持续的收入。

目前，“唯爱工坊”已成功上线“唯爱·妈妈制造——扎染专场”、“土族阿妈的盘绣潮牌”、“守望乡土真味——黑土麦田专场”等系列，并联动国内外知名设计师对东方美学进行重构，让传统技艺更时尚、更实用、更受市场的欢迎，白族阿妈的扎染、土族阿妈的盘绣这些珍贵的非遗技艺被越来越多的消费者认识并喜欢。

第9章　中国互联网未来展望

站在新的历史起点上，中国网信事业必须坚持以习近平新时代中国特色社会主义思想为指导，全面贯彻落实党的十九大精神，全面贯彻以人民为中心的发展思想，加快实施网络强国战略，大力推进数字中国建设，努力实现中国互联网全面、平衡、充分、可持续发展，让互联网更好地造福国家和人民。

9.1　更坚实的网络基础设施

继续深入实施“宽带中国”战略，加快信息基础设施升级，进一步推进固定宽带网络光纤化、移动网络宽带化、下一代互联网建设，加快推进超宽带光纤接入、5G 试验和商用、IPv6 发展，打通陆上信息丝绸之路、海上信息丝绸之路、天地一体信息网络，基本建成多网络融合发展、多空间维度一体化、综合智能化的高速、移动、安全、泛在的先进信息基础设施。继续深入开展网络提速降费行动，进一步完善电信普遍服务机制，加快农村互联网和偏远地区互联网基础设施建设，扩大光纤网、宽带网的有效覆盖，大幅提升互联网普及水平，大幅提升全民信息素质，为缩小数字鸿沟、建设智慧社会、提升全国人民共享互联网发展成果的获得感奠定坚实的网络基础设施。

9.2　更先进的网络信息技术

坚定不移实施创新驱动发展战略，瞄准世界网信技术前沿，强化基础研究，实现前瞻性研究、引领性原创成果重大突破。加强应用基础研究，实施国家网信技术重大科技项目，把更多人力、物力、财力投向核心技术研发，集合精锐力量，作出战略性安排，力争在基础技术、通用技术、非对称技术、前沿技术、颠覆性技术等方面取得显著突破，着力在网络信息核心技术领域实现从跟跑并跑到并跑领跑的重大转变。

9.3　更繁荣的数字经济

深化供给侧结构性改革，抓住新一轮社会信息化与经济全球化交汇融合发展机遇，推动互联网、大数据、人工智能与实体经济融合发展，以信息流带动技术流、资金流、人才流、物资流，促进资源配置优化，促进全要素生产率提升，升级改造商品与服务的生产、流通与销售过程，重塑业态结构与生态圈，改变经济增长方式和经济结构，激活全社会创新思维，为经济发展注入新动能、增添新动力。推动电子商务进一步创新发展，跨境电子商务快速发展，产业组织、商业模式、供应链、物流链等基于互联网的各类创新不断涌现。促进信息领域软硬件融合发展，鼓励支持智能终端、智能汽车、智能装备、智能家居、物联网、工业互联网、在线医疗、车联网等新型业态快速发展，并向深度和广度裂变。云计算和大数据领域深入发展，云计算行业巨头纷纷加快打造各自的云生态，强化对云计算行业掌控力；移动云应用加快普及，混合云技术和产品逐步成熟，工业云平台发展迅猛；以阿里云、腾讯云、华为云等为首的龙头企业将不断扩大海外业务市场，有望成为我国服务输出的重要方向。在大数据领域，国家大数据战略深入实施，数据资源开放共享进程明显加快，综合运用数据资源进行大数据挖掘分析、服务经济社会发展的能力显著提升，各领域对大数据服务需求的增强将带来产业发展的“黄金期”，大数据技术与其他新一代信息技术之间的融合创新更将成为产业发展的主基调。互联网与制造业加速融合，生产方式向柔性、智能、精细转变，智能制造加速构建新型制造体系。互联网信息技术在农业领域应用也将继续深化，进一步促进农业生产、流通组织方式的变革和营销、价值创造模式的创新，加速农业产业链的重构。共享经济将进入快速发展期，产业规模将持续快速扩大，商业模式不断创新并走向成熟。“一带一路”建设拓展数字经济国际合作新空间，“中国—东盟信息港”、网上丝路宁夏枢纽工程、中欧数字丝绸之路等平台建设体系化推进，引领带动企业加快“走出去”步伐。

9.4 更清朗的网络空间

网上网下形成同心圆，发挥网络引导舆论、反映民意的作用，把各方面的力量凝聚起来、团结起来，使全社会方方面面同心干，全国各族人民心往一处想、劲往一处使，共同为实现中华民族伟大复兴的中国梦而努力奋斗。中国将进一步加强网络内容建设，实施网络内容建设工程，推动传统媒体和新兴媒体加快融合发展，建成一批新型主流媒体，巩固壮大网上思想文化阵地；牢牢把握正确舆论导向，做强网上正面宣传，用社会主义核心价值观和优秀文明成果滋养人心、滋养社会，做到正能量充沛、主旋律高昂；发展积极健康、向上向善的网络文化，增强网络文化创新能力，推动网络文艺全面发展繁荣，打造一批具有中国气派、中国风格的网络文化品牌。中国还将继续推进依法治网，依托不断完善的互联网法律体系，加强网络空间治理，深入净化网络环境，规范网络秩序，坚决制止打击不符合人民利益的网上不法行为和非法活动，使广大网民特别是青少年享受天朗气清、生态良好的网络空间，使之成为亿万民众共同的精神家园。互联网国际传播能力将得到全面提升，网络媒体和社交平台成为讲好中国故事、传播中国声音的重要平台，推动中国文化走出去。基本走出一条政府、企业密切协作协调，齐抓共管、良性互动、合力治理的新路。

9.5　更可靠的网络安全

进一步加强顶层设计和统筹协调，围绕落实《网络安全法》《国家网络空间安全战略》，加快网络安全领域法律、法规、政策体系的完善，制定出台国家网络可信身份战略，建立健全关键信息基础设施保护、大数据安全管理制度、网络安全审查、个人数据保护等相关制度、条例和细则，打击网络违法犯罪行为和不良信息内容传播的力度不断加大，有效保障网民合法权益。网络安全相关基础性工作将得到继续深化，网络安全态势感知基础工作将进一步推进，逐步形成全天候全方位感知网络安全态势的能力，增强网络安全防御能力和威慑能力，网络安全检查持续开展，统一高效的网络安全风险报告机制、情报共享机制、研判处置机制得以建立，形成政府和企业网络安全信息共享机制，依靠大数据挖掘分析，风险防范能力大幅度提升，关键信息基础设施和重要信息系统的网络安全保障能力将进一步增强，公民个人信息的保护进一步加强。全社会网络安全意识和网络素养得到较大幅度提高，维护网络安全成为政府、企业、社会组织和广大网民共同责任，筑起全社会共同参与的网络安全防线。

9.6 更公正的国际治理

网络空间是人类共同的活动空间，各国在网络空间互联互通，利益交融，休戚与共。维护网络空间和平与安全，促进开放与合作，符合国际社会的共同利益，也是国际社会的共同责任。面向未来，中国也将以习近平总书记“四项原则”和“五点主张”为指引，更深度地参与网络空间国际交流与合作，推动根服务器布局与管理更加国际化、更具代表性、更符合全球互联网发展实际。中国将积极承担起网络大国责任，为全球网络空间发展治理提供中国方案、贡献中国智慧，充分发挥世界互联网大会的作用，推动建设新型全球数字经济发展平台、网络空间国际治理平台，深化网络空间多边、双边、多方交流与合作，携手国际社会共同构建和平、安全、开放、合作、有序的网络空间，建立多边、民主、透明的全球互联网治理体系，让互联网更安全、更充分、更平衡、可持续发展，更好地满足人们对互联网带来美好生活的需要。

展望未来，中国互联网将以更加开放的胸怀、更加包容的理念、更加创新的精神、更加务实的举措，向着网络基础设施基本普及、自主创新能力显著增强、数字经济全面发展、网络安全保障有力的目标不断前进，为夺取新时代中国特色社会主义伟大胜利，实现“两个一百年”奋斗目标和建设社会主义现代化强国目标、实现中华民族伟大复兴的中国梦，为全人类的福祉做出新的更大贡献！

附录：中国互联网发展大事记

2015 年，第二届世界互联网大会发布了《中国互联网 20 年发展报告》，对 2012 年之前的互联网发展大事进行了详尽梳理。在此基础上，我们对党的十八大以来中国互联网发展大事作了整理汇总。

2012 年

2012 年 1 月 18 日，由中国主导制定、大唐电信集团提出的 TD-LTE，被国际电信联盟确定为第四代移动通信国际标准之一。

2012 年 2 月 14 日，工业和信息化部发布《物联网“十二五”发展规划》。

2012 年 3 月 27 日，国家发展改革委等 7 部门研究并印发《关于下一代互联网“十二五”发展建设的意见》。

2012 年 4 月，人民网在国内 A 股成功上市，实现了中央重点新闻网站上市零的突破，标志着我国网络媒体建设进入新阶段。

2012 年 5 月 9 日，国务院常务会议审议通过《关于大力推进信息化发展和切实保障信息安全的若干意见》。

2012 年 7 月 9 日，国务院印发《“十二五”国家战略性新兴产业发展规划》，提出实施“宽带中国”工程。

2012 年 9 月 18 日，科技部发布《中国云科技发展“十二五”专项规划》，以加快推进云计算技术创新和产业发展。

2012 年 9 月 18 日，新兴国家互联网圆桌会议在北京举行，这是新兴国家间首次就互联网问题开展对话交流。

2012 年 11 月 1 日，在中国互联网协会组织下，12 家搜索引擎服务企业签署《互联网搜索引擎服务自律公约》。

2012 年 11 月 8 日，中国共产党第十八次全国代表大会报告指出，将促进工业化、信息化、城镇化、农业现代化同步发展，并写入《中国共产党章程》。

2012 年 12 月 5 日，首届中韩互联网圆桌会议在北京开幕。

2012 年 12 月 28 日，全国人民代表大会常务委员会审议通过《关于加强网络信息保护的决定》，要求保护个人电子信息、防范垃圾电子信息、确立网络身份管理制度并赋予了有关主管部门必要的监管权力。

2013 年

2013 年 2 月 17 日，国务院公布《关于推进物联网有序健康发展的指导意见》。国家发展改革委等部门联合印发《物联网发展专项行动计划

（2013～2015）》，推出10个物联网发展专项行动计划。

2013年6月，美国“棱镜门”事件引起国际社会对信息安全保障的高度重视，促使我国加强自主可控的信息安全建设。

2013年6月25日，在公安部指导下，阿里巴巴、腾讯、百度、新浪、盛大、网易、亚马逊中国等21家互联网企业成立“互联网反欺诈委员会”，以推进全网联合打击网络诈骗，共建交易安全生态圈。

2013年6月26日，国际互联网协会发布第二批“互联网名人堂”入选者名单，中国互联网协会理事长胡启恒院士入选并成为首个获此殊荣的中国人。

2013年7月8日，中美两国确定在中美战略安全对话框架下设立网络安全工作组，并在华盛顿举行第一次会议。

2013年7月16日，工业和信息化部公布《电信和互联网用户个人信息保护规定》。

2013年8月1日，国务院印发《“宽带中国”战略及实施方案》，指出宽带是我国经济社会发展的战略性公共基础设施，制定了2015年和2020年两阶段发展目标。

2013年8月14日，国务院印发《关于促进信息消费扩大内需的若干意见》。

2013年9月9日，最高人民法院和最高人民检察院出台《关于办理利用信息网络实施诽谤等刑事案件适用法律若干问题的解释》，自 9 月 10 日起施行。

2013 年 10 月 25 日，新版《中华人民共和国消费者权益保护法》发布，规定经营者采用网络、电视、电话、邮购等方式销售商品，消费者有权自收到商品之日起七日内退货，加强对个人信息保护以及网络交易平台的责任规定等。

2013 年 11 月 9 日，中国共产党第十八届三中全会召开。《中共中央关于全面深化改革若干重大问题的决定》指出，加大依法管理网络力度，加快完善互联网管理领导体制。

2013 年 12 月 4 日，我国正式发放首批 4G 牌照，中国移动通信集团公司、中国电信集团公司和中国联合网络通信集团有限公司获颁“LTE/第四代数字蜂窝移动通信业务（TD-LTE）”经营许可。

2013 年 12 月 26 日，工业和信息化部颁发首批“移动通信转售业务”运营试点资格，11 家民营企业获得虚拟运营商牌照。

2013 年 12 月，第二届中韩互联网圆桌会议在韩国首尔举行。

2014 年

2014 年 2 月 27 日，中共中央总书记、国家主席、中央军委主席、中央网络安全和信息化领导小组组长习近平主持召开中央网络安全和信息化领导小组第一次会议并发表重要讲话。

2014年4月，全国“扫黄打非”工作小组办公室、国家网信办、工信部、公安部发布公告，决定自2014年4月中旬至11月，在全国范围内统一开展打击网上淫秽色情信息“扫黄打非·净网2014”专项行动。

2014年6月12日，国家版权局、国家网信办、工信部、公安部四部门启动“剑网2014”，打击网络侵权盗版专项行动。

2014年7月16日，习近平主席在巴西国会发表演讲时指出，中国愿意同世界各国携手努力，本着相互尊重、相互信任的原则，深化国际合作，尊重网络主权，维护网络安全，共同构建和平、安全、开放、合作的网络空间，建立多边、民主、透明的国际互联网治理体系。

2014年7月起，中文新通用顶级域名“.公司”、“.网络”正式开放注册。

2014年8月7日，国家网信办发布《即时通信工具公众信息服务发展管理暂行规定》，对即时通信工具服务提供者、使用者的服务和使用行为进行了规范，对通过即时通信工具从事公众信息服务活动提出了明确管理要求。

2014年8月18日，中央全面深化改革领导小组第四次会议审议通过《关于推动传统媒体和新兴媒体融合发展的指导意见》。

2014年10月21日，国家新闻出版广电总局与国家网信办联合下发《关于在新闻网站核发新闻记者证的通知》，加强新闻网站编辑记者队伍建设。

2014年10月23日，中国共产党第十八届中央委员会第四次全体会议通过《中共中央关于全面推进依法治国若干重大问题的决定》，提出加强互联网领域立法，完善网络信息服务、网络安全保护、网络社会管理等方面的法律法规，依法规范网络行为。

2014 年 11 月 6 日，国家网信办、国家新闻出版广电总局在全国开展清理整治网络视频有害信息专项行动。

2014 年 11 月 19—21 日，由国家网信办和浙江省人民政府联合主办的首届世界互联网大会成功举办，国家主席习近平致贺词，李克强总理会见与会代表并座谈，马凯副总理出席大会开幕式并发表主旨演讲。

2014 年 11 月 24 日，由中央网信办等多部委联合主办的首届中国国家网络安全宣传周开幕，中央政治局常委、中央书记处书记、中央网络安全和信息化领导小组副组长刘云山同志出席启动仪式并发表重要讲话。马凯、刘奇葆、郭声琨出席。

2014 年 4 月 8 日，中国科学院计算机网络信息中心钱华林研究员入选国际互联网协会“互联网名人堂”。

2015 年

2015 年 1 月 5 日，中国人民银行印发《关于做好个人征信业务准备工作的通知》，标志着基于互联网业态的大数据征信全面开启。

2015 年 1 月 6 日，国务院印发《关于促进云计算创新发展培育信息产业新业态的意见》。

2015年2月4日，国家网信办发布《互联网用户账号名称管理规定》，即“账号十条”。

2015年3月1日，我国正式实施《宽带接入网业务开放试点方案》，宽带接入开始向民间资本开放，首批确定太原、沈阳、哈尔滨、上海、南京、杭州、宁波等16个试点城市。

2015年3月2日，中国文化网络传播研究会获民政部正式批复。

2015年3月，“互联网+”写入政府工作报告，随后国务院印发《关于积极推进“互联网+”行动的指导意见》。

2015年4月15日，全国首家大数据交易所在贵阳成立，率先推动了数据互联共享方面的探索。

2015年4月28日，国家网信办出台《互联网新闻信息服务单位约谈工作规定》，即“约谈十条”。

2015年4月29日，中国文化网络传播研究会成立。

2015年5月4日，国务院印发《关于大力发展电子商务加快培育经济新动力的意见》。

2015年5月8日，国务院出台《中国制造2025》，部署全面推进实施制造强国战略。

2015年8月31日，国务院印发《促进大数据发展行动纲要》，全面推进中国大数据发展和应用，加快建设数据强国。

2015 年 6 月 8 日，中宣部与中央网信办共同召开网络公益活动推进会，参会互联网站和互联网企业签署《共同推进网络公益倡议书》。

2015 年 6 月 12 日，国家网信办举办“网络诚信伴我行”为主题的首届网络诚信宣传日活动，提出“三实三主动”，大力倡导网站诚信办网、网民诚信用网的理念。

2015 年 6 月 30 日，阿里巴巴董事局主席马云当选为联盟理事会联合主席。

2015 年 7 月 1 日，《中华人民共和国国家安全法》正式实施，提出加强网络管理，防范、制止和依法惩治网络攻击、网络入侵、网络窃密、散布违法有害信息等网络违法犯罪行为，维护国家网络空间主权、安全和发展利益。

2015 年 7 月 6 日，第一届中欧数字合作圆桌会议在比利时召开。

2015 年 7 月 18 日，中国人民银行等十部委联合发布了《关于促进互联网金融健康发展的指导意见》。

2015 年 8 月 3 日，中国互联网发展基金会在北京正式挂牌，标志着中国也是全球首家互联网领域的公募基金会正式成立。

2015 年 9 月 13 日，由国家网信办、国家发改委与广西壮族自治区政府共同主办“中国-东盟信息港论坛”在南宁举行。

2015 年 9 月 23 日，第八届中美互联网论坛在美国举行，国家主席习近平会见出席论坛代表并发表讲话。

2015 年 10 月 20 日，2015 年世界物联网大会在江苏无锡举行。

2015 年 10 月 26 日，中共中央十八届五中全会召开，通过的《中共中央关于制定国民经济和社会发展第十三个五年规划的建议》明确提出实施国家安全战略、网络强国战略、国家大数据战略、军民融合发展战略，加强网上思想文化阵地建设，推动传统媒体和新兴媒体融合发展、加强国际传播能力建设等重大任务。

2015 年 11 月 1 日，《刑法修正案（九)》正式颁布实施，明确了网络服务提供者履行信息网络安全管理的义务，加大对信息网络犯罪的刑罚力度，加强对公民个人信息的保护。

2015 年 11 月 6 日，首批新闻网站记者证发放，人民网、新华网、光明网等 14 家中央主要新闻网站首批 594 名记者获发新闻记者证，成为首批“持证上岗”网络媒体记者。

2015 年 11 月 23 日，世界机器人大会在北京举行，国家主席习近平向大会致贺信。

2015 年 12 月 1 日，首次中美打击网络犯罪及相关事项高级别对话在美国华盛顿举行，国务委员郭声琨与美国司法部部长林奇、国土安全部部长约翰逊共同主持。

2015 年 12 月 16 日，第二届世界互联网大会在浙江乌镇开幕，国家主席习近平出席大会并发表主旨演讲，提出推进全球互联网治理体系变革“四项原则”和构建网络空间命运共同体“五点主张”。

2016 年

2016 年 2 月 4 日，国家新闻出版广电总局、工信部发布《网络出版服务管理规定》。

2016 年 3 月 13 日，中央网信办启动争做中国好网民工程，团结凝聚网民力量，共建清朗网络空间。

2016 年 3 月 25 日，中国网络空间安全协会成立。

2016 年 4 月 19 日，中共中央总书记、国家主席、中央军委主席、中央网络安全和信息化领导小组组长习近平主持召开网络安全和信息化工作座谈会并发表重要讲话。中共中央政治局常委、中央网络安全和信息化领导小组副组长李克强、刘云山出席座谈会。马凯、王沪宁、刘奇葆、范长龙、孟建柱、栗战书、杨洁篪、周小川出席座谈会。

2016 年 5 月 5 日，工信部向中国广播电视网络有限公司颁发《基础电信业务经营许可证》，批准中国广播电视网络有限公司在全国范围内经营互联网国内数据传送业务、国内通信设施服务业务。

2016 年 5 月 18 日，国家发展改革委、科技部、工信部、中央网信办联合制订《“互联网+”人工智能三年行动实施方案》，在国家战略层面将人工智能产业作为发展重点。

2016 年 6 月 14 日，第二次中美打击网络犯罪及相关事项高级别联合对话在北京举行，中美双方通过《中美打击网络犯罪及相关事项热线机制运作方案》，并联合发表成果清单。

2016 年 6 月 25 日，中俄在北京签署《关于协作推进信息网络空间发展的联合声明》，致力于两国推进信息网络空间发展。

2016 年 6 月 25 日，国家网信办发布《互联网信息搜索服务管理规定》。

2016 年 6 月 28 日，国家网信办发布《移动互联网应用程序信息服务管理规定》。

2016 年 7 月 4 日，国家工商总局发布《互联网广告管理暂行办法》。

2016 年 7 月 27 日，中共中央办公厅、国务院办公厅发布《国家信息化发展战略纲要》。

2016 年 9 月 4 日至 5 日，二十国集团（G20）领导人峰会在杭州举行，峰会通过了《G20 数字经济发展与合作倡议》，是全球首个由多国领导人共同签署的数字经济政策文件。

2016 年 9 月 13 日，一审判决深圳市快播科技有限公司及被告人犯传播淫秽物品牟利罪。

2016 年 11 月 1 日，《网络预约出租汽车经营服务管理暂行办法》正式施行。

2016 年 11 月 4 日，国家网信办发布《互联网直播服务管理规定》。

2016 年 11 月 7 日，十二届全国人大常委会第二十四次会议表决通过《中华人民共和国网络安全法》，这是我国第一部全面规范网络空间安全管理方面问题的基础性法律，是我国网络空间法治建设的重要里程碑。

2016 年 11 月 16—18 日，第三届世界互联网大会在浙江乌镇举办，中共中央总书记、国家主席习近平在会上以视频方式发表重要讲话，中共中央政治局常委、中央书记处书记刘云山出席大会并发表致辞。

2016 年 12 月 27 日，国家网信办发布《国家网络空间安全战略》，是我国首部关于国家网络安全工作的纲领性文件。

2017 年

2017 年 1 月 10 日，国家发展改革委、中央网信办、商务部等部门印发《促进电子商务发展三年行动实施方案(2016—2018)》。

2017 年 1 月 13 日，中国人民银行发布《关于实施支付机构客户备付金集中存管有关事项的通知》，标志着人民银行对移动支付机构客户备付金实施集中存管机制的建立。

2017 年 1 月 15 日，中共中央办公厅、国务院办公厅印发《关于促进移动互联网健康有序发展的意见》，进一步明确了推动和规范移动互联网发展的政策方向和原则。

2017 年 1 月 16 日，国家发展改革委、商务部等 9 部委印发《关于全面加强电子商务领域诚信建设的指导意见》，明确要加强电子商务全流程信用建设，强化消费者权益保障措施。

2017 年 1 月 17 日，工信部发布《信息通信行业发展规划（2016—2020 年）》。

2017 年 1 月 22 日，由国家网信办和财政部共同发起的中国互联网投资基金在京成立，基金总规模为 1 000 亿元，首期募集资金 300 亿元。

2017 年 1 月 22 日，工信部印发《关于清理规范互联网网络接入服务市场的通知》，旨在进一步规范市场秩序、强化网络信息安全管理，促进互联网行业健康有序发展。

2017 年 2 月 22 日，银监会发布《网络借贷资金存管业务指引》，明确网贷资金存管业务应遵循的基本规则和实施标准，鼓励网贷机构与商业银行按照平等自愿互利互惠的市场化原则开展业务。

2017 年 2 月 24 日，招行宣布通过区块链技术实现直联跨境支付应用，标志国内首个区块链跨境领域项目落地应用，在国内区块链金融领域具有里程碑意义。5 月 16 日，在杭州举行的“区块链技术和应用峰会暨首届中国区块链开发大赛”上，中国区块链技术和产业发展论坛公布了《区块链和分布式账本技术参考架构》标准，这是中国首个政府指导下的区块链基础标准。

2017 年 3 月 1 日，外交部和国家网信办共同发布《网络空间国际合作战略》。

2017年5月2日，国家网信办公布《互联网新闻信息服务管理规定》，对互联网新闻信息服务许可管理、网信管理体制、互联网新闻信息服务提供者主体责任等作出规定。

2017年5月2日，国家网信办发布《互联网信息内容管理行政执法程序规定》，自2017年6月1日起施行。

2017年5月3日，由中国科技大学、中国科学院-阿里巴巴量子计算实验室、浙江大学、中科院物理所等协同完成研发的世界首台超越早期经典计算机的光量子计算机在中国诞生。

2017年5月14日，国家主席习近平出席“一带一路”国际合作高峰论坛开幕式，并发表题为《携手推进“一带一路”建设》的主旨演讲，指出“我们要坚持创新驱动发展，加强在数字经济、人工智能、纳米技术、量子计算机等前沿领域合作，推动大数据、云计算、智慧城市建设，连接成21世纪的数字丝绸之路”。

2017年5月22日，国家网信办公布《互联网新闻信息服务许可管理实施细则》，自2017年6月1日起施行。

2017年6月1日，《中华人民共和国网络安全法》正式施行。

2017年6月26日，中央全面深化改革领导小组第三十六次会议审议通过《关于设立杭州互联网法院的方案》，8月18日杭州互联网法院挂牌成立，成为全国第一家互联网法院。

2017年6月28日，银监会、教育部、人力资源和社会保障部联合下发《关于进一步加强校园贷规范管理工作的通知》，要求“一律暂停网贷机构开展在校大学生网贷业务，逐步消化存量业务”。

2017 年 7 月 3 日，国家发展改革委等八部门联合印发《关于促进分享经济发展的指导性意见》。

2017 年 7 月 8 日，国务院印发《新一代人工智能发展规划》，第一次将人工智能发展放在国家战略层面。

2017 年 8 月 25 日，国家网信办发布《互联网论坛社区服务管理规定》《互联网跟帖评论服务管理规定》，自 10 月 1 日起施行。

2017 年 9 月 4 日，中国人民银行与国家网信办等七部门联合发布公告，要求即日起各类代币发行融资活动应立即停止，已完成代币发行融资的组织应作出清退安排，合理保护投资者权益，妥善处置风险，并表示将依法查处拒不停止的代币发行融资活动以及已完成代币发行融资项目中的违法违规行为。

2017 年 9 月 7 日，国家网信办发布《互联网用户公众账号信息服务管理规定》《互联网群组信息服务管理规定》，自 10 月 8 日起施行。

2017 年 9 月 18 日，中国工程院院士吴建平入选国际互联网协会“互联网名人堂”。

2017 年 9 月 29 日,世界首条量子保密通信网络“京沪干线”正式开通。

后　记

党的十九大将习近平新时代中国特色社会主义思想确立为党必须长期坚持的指导思想，勾画了党和国家事业发展的宏伟蓝图，特别是对推动互联网、大数据、人工智能、共享经济发展，建设网络强国、数字中国、智慧社会等作出重大战略部署，为中国互联网发展提供了根本遵循与行动指南。新时代开辟新征程，新时代肩负新使命。我们希望通过《中国互联网发展报告 2017》（以下简称《报告》）的编撰，深入宣传阐释习近平新时代中国特色社会主义思想特别是网络强国战略思想，全面展现我国互联网发展成就，系统总结互联网发展治理的中国经验，科学展望中国互联网的发展前景，更好推动我国互联网发展。我们也希望通过问道中国互联网发展，为世界各国互联网发展治理提供中国经验、贡献中国智慧。

《报告》的编撰得到了中央网信办的指导和支持。网信办领导对《报告》给予了具体指导，网信办内部各局、各事业单位对报告编写工作特别是相关数据和素材方面给予了大力支持。在《报告》编撰过程中，中国网络空间研究院牵头，组织中国互联网络信息中心、浙江省工业和信息化研究院、中国信通院、工信部电子一所、工信部安全中心、国家信息中心、赛迪智库、清华大学国家治理研究院、北京大学互联网发展研究中心、复旦大学网络空间治理研究中心和互联网实验室等智库机构共同参与研究编写工作。

《报告》的顺利出版也离不开社会各界的大力支持和帮助，但鉴于研究经验、能力水平和编写时间有限，《报告》难免存在不足之处。为此，我们殷切地希望国内外政府部门、国际组织、科研院所、互联网企业、社会团体及各界人士对《报告》提出宝贵的修改意见和建议，提供详尽的素材资料，进一步完善报告质量，为我国互联网的发展提供更加坚实的智力支持。

中国网络空间研究院

2017 年 12 月

举报电话：（010）88254396；（010）88258888

传　　真：（010）88254397

E-mail:　　dbqq@phei.com.cn

通信地址：北京市万寿路 173 信箱

　　　　　电子工业出版社总编办公室

邮　　编：100036